EVOLUTION
THE FOSSILS SAY
NO!

Public School Edition

DUANE T. GISH, Ph.D.

CREATION-LIFE PUBLISHERS
San Diego, California

EVOLUTION? The Fossils Say NO!
Public School Edition

CREATION-LIFE PUBLISHERS
P.O. Box 15666
San Diego, California 92115

Library of Congress Catalog Card Number 78-52337
ISBN 0-89051-046-6

Library of Congress Cataloging in Publication Data
Gish, Duane Tolbert, 1921-
Evolution? The fossils say NO! (Public school edition)
1. Evolution. 2. Paleontology. I. Title.
QH369.G54 575 78-52337
ISBN 0-89051-046-6

Printed in United States of America

Printed by
El Camino Press
La Verne, California

ABOUT THE AUTHOR

DUANE T. GISH, Ph.D. (Biochemistry, University of California, Berkeley), is Associate Director of the Institute for Creation Research and Professor of Natural Science at Christian Heritage College, San Diego, California. He spent 18 years in biochemical and biomedical research at Cornell University Medical College, the Virus Laboratory of the University of California, Berkeley, and The Upjohn Company, Kalamazoo, Michigan. He is the author or co-author of numerous technical articles in his field and a well-known author and lecturer on creation/evolution.

DEDICATION

This book is dedicated to Sidney J. Jansma, one of the outstanding lay leaders of America in the cause of scientific creationism.

PREFACE

This book constitutes one of the most devastating critiques of the evolutionary philosophy one could find. It goes right to the stronghold of the supposed scientific evidence for evolution and demolishes its central bastion.

The fossil record must provide the critical evidence for or against evolution, since no other scientific evidence can possibly throw light on the actual history of living things. All other evidences are circumstantial and can be more effectively explained in terms of the creation model. The time scale of human observation is far too short to permit documentation of real evolutionary change from lower to higher kinds of organisms at the present time. The vital question, therefore, is: "Does the record of past ages, now preserved in the form of fossils, show that such changes have occurred?" The answer, unequivocally, is: "The fossils say no!" There has been no evolution in the past, any more than in the present. This important fact is conculsively demonstrated and documented by Dr. Gish in this book.

Dr. Duane T. Gish is a careful scientist of impeccable academic credentials. He has successfully defended creationism before numerous university and scientific audiences and in formal debates with many of the nation's leading evolutionary scientists.

This book has gone through several printings in its first edition and has already been eminently successful in its mission of convincing men of the truth of creationism. In this new enlarged edition, it is still more convincing, and will, no doubt, have a greater acceptance than ever before. Anyone who reads this book and who then still rejects creationism in favor of evolutionism must at least acknowledge that he *believes* in evolution in *spite* of the massive witness of the fossil record *against* it!

Henry M. Morris, Ph.D., Director
Institute for Creation Research
San Diego, California

CONTENTS

Chapter I

EVOLUTION— A PHILOSOPHY, NOT A SCIENCE

The general theory of organic evolution, or the evolution model, is the theory that all living things have arisen by a materialistic evolutionary process from a single source which itself arose by a similar process from a dead, inanimate world. This theory may also be called the molecule-to-man theory of evolution.

The creation model, on the other hand, postulates that all basic animal and plant types (the created kinds) were brought into existence by acts of a supernatural Creator using special processes which are not operative today.

Most scientists accept evolution, not as a theory, but as an established fact. The late Theodosius Dobzhansky, geneticist and widely-known evolutionist, formerly Professor of Zoology at Columbia University and visiting Professor at the University of California, Davis, has said that, "The occurrence of the evolution of life in the history of the earth is established about as well as events not witnessed by human observers can be."[1] Richard B.

Goldschmidt, a Professor at the University of California before his death, has stated that, "Evolution of the animal and plant world is considered by all those entitled to judgment to be a fact for which no further proof is needed."[2] Almost all science books and school and university texts present evolution as an established fact. These considerations alone convince many people that molecule-to-man evolution has actually occurred.

The proponents of evolution theory adamantly insist that special creation be excluded from any possible consideration as an explanation for origins on the basis that it does not qualify as a scientific theory. On the other hand, they would view as unthinkable the consideration of evolution as anything less than pure science. In fact, as already mentioned, most evolutionists insist that evolution must no longer be thought of as a theory, but must be considered to be a fact. In spite of this attitude, however, not only is there a wealth of scientific support for rejecting evolution as a fact, but evolution does not even qualify as a scientific theory according to a strict definition of the latter.

What criteria must be met for a theory to be considered as scientific in the usually accepted sense? George Gaylord Simpson has stated that, "It is inherent in any definition of science that statements that cannot be checked by observation are not really about anything . . . or at the very least, they are not science."[3] A definition of science given by the Oxford Dictionary is:

> A branch of study which is concerned either with a connected body of *demonstrated truths* or with *observed facts* systematically classified and more or less colligated by being brought under

> general laws, and which includes trustworthy methods for the discovery of new truth within its own domain. (Emphasis added.)

Thus, for a theory to qualify as a scientific theory, it must be supported by events, processes, or properties which can be observed, and the theory must be useful in predicting the outcome of future natural phenomena or laboratory experiments. An additional limitation usually imposed is that the theory must be capable of falsification. That is, it must be possible to conceive some experiment, the failure of which would disprove the theory.

It is on the basis of such criteria that most evolutionists insist that creation be refused consideration as a possible explanation for origins. Creation has not been witnessed by human observers, it cannot be tested experimentally, and as a theory it is nonfalsifiable.

The general theory of evolution also fails to meet all three of these criteria, however. It is obvious, for example, that no one observed the origin of the universe, the origin of life, the conversion of a fish into an amphibian, or an ape into a man. No one, as a matter of fact, has ever observed the origin of a species by naturally occurring processes. Evolution has been *postulated*, but it has never been *observed*.

This has been affirmed by both Dobzhansky and Goldschmidt, whom, as it has been noted, are wholly committed to faith in evolution. In the quotation cited earlier in this chapter, Dobzhansky clearly states that *evolution has not been witnessed by human observers*.

Goldschmidt, after outlining his postulated systemic mutation, or "hopeful monster," mechanism for evolution, stated:

> Such an assumption is violently opposed by the majority of geneticists, who claim that the facts found on the subspecific level must apply also to the higher categories. Incessant repetition of this *unproved claim*, glossing lightly over the difficulties, and the assumption of an arrogant attitude toward those who are not so easily swayed by fashions in science, are considered to afford scientific proof of the doctrine. It is true that nobody thus far has produced a new species or genus, etc., by macromutation. *It is equally true that nobody has produced even a species by the selection of micromutations.*[4] (Emphasis added.)

Later on in this same paper, he stated, "Neither has anyone witnessed the production of a new specimen of a higher taxonomic category by selection of micromutants."[5] Goldschmidt has thus affirmed that, in the molecules-to-man context, only the most trivial change, or that at the subspecies level, has actually ever been observed.

Since evolution has not been observed in nature, and even a species cannot be produced by the selection of mutants, it is apparent that evolution is not subject to experimental test. This was admitted by Dobzhansky when he said:

> These evolutionary happenings are unique, unrepeatable, and irreversible. It is as impossible to turn a land vertebrate into a fish as it is to effect the reverse transformation. The applicability of the experimental method to the study of such

> unique historical processes is severely restricted before all else by the time intervals involved, which far exceed the lifetime of any human experimenter. And yet it is just such *impossibility* that is demanded by anti-evolutionists when they ask for "proofs" of evolution which they would magnanimously accept as satisfactory.[6] (Emphasis added.)

Dobzhansky thus states that the applicability of the experimental method to evolution is an "impossibility." One reason given by Dobzhansky and other evolutionists for rejecting creation as a possible explanation for origins is because it is not subject to the experimental method. At the same time, however, they consider it wholly unreasonable for creationists to place the same demand on evolution theory!

It can be seen that evolutionists seek to excuse the fact that evolution cannot be observed or tested experimentally on the basis that real evolutionary events require great lengths of time for their consummation. Yes, it is true that the evolutionary process postulated would require more time than we have available for human observation. But then, evolution can never be more than just a postulate.

Macbeth, who is by no means a creationist, has flatly stated that "Darwinism is not science."[7] Birch and Ehrlich state that the theory of evolution ". . . is 'outside of empirical science' but not necessarily false. No one can think of ways in which to test it."[8]

After stating that the neo-Darwinian theory of evolution is based on axioms (concepts that can be neither proved nor tested), Harris proclaims:

> . . . the axiomatic nature of the neo-Darwinian theory places the debate between evolutionists and creationists in a new perspective. Evolutionists have often challenged creationists to provide experimental proof that species have been fashioned de novo. Creationists have often demanded that evolutionists show how chance mutations can lead to adaptability, or to explain why natural selection has favored some species but not others with special adaptations, or why natural selection allows apparently detrimental organs to persist. We may now recognize that neither challenge is fair. If the neo-Darwinian theory is axiomatic, it is not valid for creationists to demand proof of the axioms, and it is not valid for evolutionists to dismiss special creation as unproved so long as it is stated as an axiom.[9]

Matthews, British biologist and evolutionist, in his introduction to a 1971 publication of Darwin's *Origin of Species*, says:

> The fact of evolution is the backbone of biology, and biology is thus in the peculiar position of being a science founded on an unproved theory — is it then a science or faith? Belief in the theory of evolution is thus exactly parallel to belief in special creation — both are concepts which believers know to be true but neither, up to the present, has been capable of proof.[10]

While evolutionists deny the miraculous in the origin of living things, the evolutionary process, given enough time, supposedly produces miracles. Thus,

$$\text{FROG} \xrightarrow{t\ =\ \text{instantaneous}} \text{PRINCE} = \text{NURSERY TALE}$$

but

$$\text{FROG} \xrightarrow{t\ =\ 300\ \text{million years}} \text{PRINCE} = \text{SCIENCE}$$

Furthermore, the architects of the modern synthetic theory of evolution have so skillfully constructed their theory that it is not capable of falsification. The theory is so plastic that it is capable of explaining anything. This is the complaint of Olson[11] and of several participants in the Wistar Institute symposium on mathematical challenges to the neo-Darwinian interpretation of evolution.[12]

Eden, one of the mathematicians, put it this way, with reference to falsifiability:

> This cannot be done in evolution, taking it in its broad sense, and this is really all I meant when I called it tautologous in the first place. It can, indeed, explain anything. You may be ingenious or not in proposing a mechanism which looks plausible to human beings and mechanisms which are consistent with other mechanisms which you have discovered, but it is still an unfalsifiable theory.[13]

In addition to scientists who are creationists, a growing number of other scientists have expressed doubts that modern evolution theory could explain more than trivial change. Eden became so discouraged after computerized calculations showed that the probability of certain evolutionary changes occurring (according to mechanisms postulated by modern evolutionists) was essentially zero that he proclaimed, ". . . an adequate scientific theory of evolution must await the discovery and elucidation of new natural laws — physical, physicochemical, and biological."[14] Salisbury has similarly stated his doubts based on probabilistic considerations.[15]

The attack on modern formulations of the theory by French scientists has been intense in recent years. A review of the French situation stated:

> This year saw the controversy rapidly growing, until recently it culminated in the title "Should We Burn Darwin?" spread over two pages of the magazine *Science et Vie*. The article, by the science writer Aime Michel, was based on the author's interviews with such specialists as Mrs. Andree Tetry, professor at the famous *Ecole des Hautes Etudes*, and a world authority on problems of evolution; Professor Rene Chauvin, and other noted French biologists; and on his thorough study of some 600 pages of biological data collected, in collaboration with Mrs. Tetry, by the late Michael Cuenot, a biologist of international fame. Aime Michel's conclusion is significant: "The classical theory of evolution in its strict sense belongs to the past. Even if they do not publicly take a definite stand, almost all French specialists hold today strong mental reservations as to the validity of natural selection."[16]

E. C. Olson, one of the speakers at the Darwinian Centennial Celebraton at Chicago, made the following statement on that occasion:

> There exists, as well, a generally silent

> group of students engaged in biological pursuits who tend to disagree with much of the current thought, but say and write little because they are not particularly interested, do not see that controversy over evolution is of any particular importance, or are so strongly in disagreement that it seems futile to undertake the monumental task of controverting the immense body of information and theory that exists in the formulation of modern thinking. It is, of course, difficult to judge the size and composition of this silent segment, but there is no doubt that the numbers are not inconsiderable.[17]

Fothergill refers to what he calls "the paucity of evolutionary theory as a whole."[18] Erhlich and Holm have stated their reservations in the following way:

> Finally, consider the third question posed earlier: "What accounts for the observed patterns in nature?" It has become fashionable to regard modern evolutionary theory as the *only* possible explanation of these patterns rather than just the best explanation that has been developed so far. It is conceivable, even likely, that what one might facetiously call a non-Euclidean theory of evolution lies over the horizon. Perpetuation of today's theory as dogma will not encourage progress toward more satisfactory explanations of observed phenomena.[19]

Sometimes the attacks are openly critical, such as Danson's letter which appeared recently in *New Scientist* and stated in part:

> . . . the Theory of Evolution is no longer with us, because neo-Darwinism is now acknowledged as being unable to explain anything more than trivial change, and in default of some other theory we have none . . . despite the hostility of the witness provided by the fossil record, despite the innumerable difficulties, and despite the lack of even a credible theory, evolution survives . . . . Can there be any other area of science, for instance, in which a concept as intellectually barren as embryonic recapitulation could be used as evidence for a theory?[20]

Macbeth has recently published an especially incisive criticism of evolution theory.[21] He points out that although evolutionists have abandoned classical Darwinism, the modern synthetic theory they have proposed as a substitute is equally inadequate to explain progressive change as the result of natural selection, and, as a matter of fact, they cannot even define natural selection in nontautologous terms. Inadequacies of the present theory and failure of the fossil record to substantiate its predictions leave macroevolution, and even microevolution, intractable mysteries according to Macbeth. Macbeth suggests that no theory at all may be preferable to the existing one.

In a recent book,[22] Pierre Grassé, one of France's best-known scientists, has severely criticized modern evolution theory. Dobzhansky, in his review[23] of that book, stated:

> The book of Pierre P. Grassé is a frontal attack on all kinds of "Darwinism." Its

> purpose is "to destroy the myth of evolution, as a simple, understood, and explained phenomenon," and to show that evolution is a mystery about which little is, and perhaps can be, known. Now one can disagree with Grassé but not ignore him. He is the most distinguished of French zoologists, the editor of the 28 volumes of *Traité de Zoologie*, author of numerous original investigations, and ex-president of the Academie des Sciences. His knowledge of the living world is encyclopedic.

Grassé ends his book with the sentence, "It is possible that in this domain biology, impotent, yields the floor to metaphysics."

In view of the above, it is incredible that most leading scientists dogmatically insist that the molecules-to-man evolution theory be taught as a fact to the exclusion of all other postulates. Evolution in this broad sense is unproven and unprovable and thus cannot be considered as fact. It is not subject to test by the ordinary methods of experimental science — observation and falsification. It thus does not, in a strict sense, even qualify as a scientific theory. It is a postulate and may serve as a model within which attempts may be made to explain and correlate the evidence from the historical record, that is, the fossil record, and to make predictions concerning the nature of future discoveries.

Creation is, of course, unproven and unprovable by the methods of experimental science. Neither can it qualify, according to the above criteria, as a scientific theory, since creation would have been unobservable and would as a theory be nonfalsifiable. In the scientific realm, creation is, therefore, as is evolution, a postulate which may serve as a model to explain and correlate the evidence related to origins. Creation is, in this sense, no more religious nor less scientific than evolution. In fact to many well-informed scientists, creation seems to be far superior to the evolution model as an explanation for origins.

It is often stated that there are no reputable scientists who do not accept the theory of evolution. This is just one more false argument used to win converts to the theory. While it is true that creationists among scientists definitely constitute a minority, there are *many* creation scientists, and their number is growing. Among these may be numbered such well-established scientists as Dr. A. E. Wilder-Smith, Professor of Pharmacology, of Boggern, Switzerland, and author or coauthor of more than 50 technical publications; Dr. W. R. Thompson, world-famous biologist and former Director of the Commonwealth Institute of Biological Control of Canada; Dr. Melvin A. Cook, winner of the 1968 E. G. Murphee Award in Industrial and Engineering Chemistry from the American Chemical Society and also winner of the Nobel Nitro Award, now president of the Ireco Chemical Company, Salt Lake City; Dr. Henry M. Morris, for 13 years Professor of Hydraulic Engineering and Head of the Civil Engineering Department at Virginia Polytechnic Institute and University, one of the largest in the U.S., now Director of the Institute for Creation Research and Academic Vice-President of Christian Heritage College, San Diego; Dr. Walter Lammerts, geneticist and famous plant-breeder; Dr. Frank Marsh, Professor of Biology at

Andrews University until his retirement; the late Dr. J. J. Duyvene De Wit, Professor of Zoology at the University of the Orange Free State, South Africa, at the time of his death; and Dr. Thomas G. Barnes, Professor of Physics at the University of Texas, El Paso.

The Creation Research Society, a recently formed organization of Christian men and women of science, all of whom hold advanced degrees and are fully committed to the acceptance of creationism as opposed to evolution, now numbers over 600 in membership.[24] There is yet a vastly larger number of scientists who do not accept the theory but choose to remain silent for a variety of reasons.

Why have most scientists accepted the theory of evolution? Is the evidence really that convincing? This seems to be the clear implication. On the other hand, is it possible for that many scientists to be wrong? The answer is an emphatic "YES!" Consider for a moment some historical examples. For centuries the accepted scientific view was that all planets revolved around the earth. This was the Ptolemaic geocentric theory of the universe. Only after a prolonged and bitter controversy did the efforts of Copernicus, Galileo, and others succeed in convincing the scientific world that the Ptolemaic system was wrong and that Copernicus was right in his contention that the planets in the solar system revolved around the sun.

At one time most people with scientific training who rejected creation accepted as fact the idea that life spontaneously arose from nonlife. Thus, frogs supposedly spontaneously arose from swamps, decaying matter generated flies, and rats were brought to life out of matter found in debris, etc. A series of carefully designed and executed experiments by Redi, Spallanzani, and Pasteur spanning 200 years were required to put to rest the theory of the spontaneous generation of life.

In recent times, a theory dealing with weak interaction of atomic particles became so widely accepted by physicists that it won the status of a law, the Law of Parity. During the 1950's, two brilliant Chinese-American scientists performed a series of experiments that disproved the theory and deposed the "Law."

In all of the above examples, the vast majority of scientists was wrong and a small minority was right. No doubt, strong preconceived ideas and prejudices were powerful factors in accounting for the fact that scientists were reluctant to give up the geocentric theory of the universe and the theory of the spontaneous generation of life.

The effects of prejudice and preconceived ideas are of overwhelming importance in the acceptance of the theory of evolution. The reason most scientists accept evolution has nothing to do, primarily, with the evidence. The reason that most scientists accept the theory of evolution is that *most scientists prefer to believe a materialistic, naturalistic explanation for the origin of all living things*. Watson, for example, has referred to the theory of evolution as "a theory universally accepted not because it can be proved by logically coherent evidence to be true, but because the only alternative, special creation, is clearly incredible."[25] That this is the philosophy held by most biologists has been recently emphasized by Dobzhansky. In his review of Monod's book, *Chance and Necessity*, Dobzhansky said, "He has stated with admirable clarity, and eloquence often verging on pathos, *the mechanistic materialist philosophy shared by most of the present 'establishment' in the biological sciences.*"[26] (Emphasis added.)

Sir Julian Huxley, British evolutionist and grandson of Thomas Huxley, one of Darwin's strongest supporters when he first published his theory, has said that "Gods are peripheral phenomena produced by evolution."[27] What Huxley meant was that the idea of God merely evolved as man evolved from lower animals. Huxley would like to establish a humanistic religion based on evolution. Humanism has been defined as "the belief that man shapes his own destiny. It is a constructive philosophy, a *nontheistic religion*, a way of life."[28] (Emphasis added.) This same publication quotes Huxley as saying:

> I use the word "Humanist" to mean someone who believes that man is just as much a natural phenomenon as an animal or plant; that his body, mind, and soul were not supernaturally created but are products of *evolution*, and that he is not under the control or guidance of any supernatural being or beings, but has to rely on himself and his own powers.

The inseparable link between this nontheistic humanistic religion and belief in evolution is evident.

Dr. George Gaylord Simpson, Professor of Vertebrate Paleontology at Harvard University until his retirement and one of the world's best-known evolutionists, has said that the Christian faith, which he calls the "higher superstition" (in contrast to the "lower superstition" of pagan tribes of South America and Africa), is intellectually unacceptable.[29] Simpson concludes his book, *Life of the Past*,[30] with what Sir Julian Huxley has called "a splendid assertion of the evolutionist view of man."[31] Simpson writes:

> Man stands alone in the universe, a unique product of a long, unconscious, impersonal, material process with unique understanding and potentialities. These he owes to no one but himself, and it is to himself that he is responsible. He is not the creature of uncontrollable and undeterminable forces, but his own master. He can and must decide and manage his own destiny.

Thus, according to Simpson, man is alone in the Universe (there is no God), he is the result of an impersonal, unconscious process (no one directed his origin or creation), and he is his own master and must manage his own destiny (there is no God who is man's Lord and Master and who determines man's destiny). That, according to Simpson and Huxley, is *the evolutionist's view of man.*

No doubt a large majority of the scientific community embraces the mechanistic materialistic philosophy of Simpson, Huxley, and Monod. Many of these men are highly intelligent, and they have woven the fabric of evolution theory in an ingenious fashion. They have then combined this evolution theory with humanistic philosophy and have clothed the whole with the term, "science." The product, a nontheistic religion, with evolutionary philosophy as its creed under the guise of "science," is being taught in most public schools, colleges, and universities of the United States. It has become our unofficial state-sanctioned religion.

Not all evolutionists are materialistic atheists or agnostics. Many evolutionists believe in God. They believe that evolution was God's method of creation, that God initiated the process at the molecular level and then allowed it to follow its natural course. The scientific evidence, however, speaks just as strongly against theistic evolution as it does against any other form of evolution.

It is apparent that acceptance of creation requires an important element of faith. Of course, belief in evolution also requires a vitally important element of faith. According to one of the most popular theories on the origin of the universe, all energy and matter of the universe was once contained in a plasma ball of electrons, protons, and neutrons (how it got there, no one has the faintest notion). This huge cosmic egg then exploded—and here we are today, several billion years later, human beings with a three-pound brain composed of 12 billion brain cells connected to about 10 thousand other brain cells in the most complicated arrangement of matter known to man (there are 120 trillion connections in the human brain!).

If this is true, then what we are and how we came to be were due solely to the properties inherent in electrons, protons, and neutrons. To believe *this* obviously requires a tremendous exercise of faith. Evolution theory is indeed no less religious nor more scientific than creation.

The question is, then, who has more evidence for his faith, the creationist or the evolutionist? The scientific case for special creation, as we will show in the following pages, is much stronger than the case for evolution. The more I study and the more I learn, the more I become convinced that evolution is a false theory and that special creation offers a much more satisfactory interpretive framework for correlating and explaining the scientific evidence related to origins.

REFERENCES

1. T. Dobzhansky, *Science*, Vol. 127, p. 1091 (1958).
2. R. B. Goldschmidt, *American Scientist*, Vol. 40, p. 84 (1952).
3. G. G. Simpson, *Science*, Vol. 143, p. 769 (1964).
4. R. B. Goldschmidt, Ref. 2, p. 94.
5. R. B. Goldschmidt, Ref. 2, p. 97.
6. T. Dobzhansky, *American Scientist*, Vol. 45, p. 388 (1957).
7. N. Macbeth, *American Biology Teacher*, November 1976, p. 496.
8. L. C. Birch and P. R. Ehrlich, *Nature*, Vol. 214, p. 349 (1967).
9. C. Leon Harris, *Perspectives in Biology and Medicine*, Winter 1975, p. 183.
10. L. Harrison Matthews, Introduction to *The Origin of Species*, C. Darwin, reprinted by J. M. Dent and Sons, Ltd., London, 1971, p. XI.
11. E. C. Olson, in *Evolution After Darwin*, Vol. 1: *The Evolution of Life*, ed. by Sol Tax, University of Chicago Press, Chicago, 1960.
12. P. S. Moorhead and M. M. Kaplan, eds., *Mathematical Challenges to the Neo-Darwinian Interpretation of Evolution*, Wistar Institute Press, Philadelphia, 1967, pp. 47, 64, 67, 71.
13. M. Eden, Ref. 12, p. 71.
14. M. Eden, Ref. 12, p. 109.

15. F. Salisbury, *Nature*, Vol. 224, p. 342 (1969); *American Biology Teacher*, Vol. 33, p. 335 (1971).
16. L. Litynski, *Science Digest*, Vol. 50, p. 61 (1961).
17. E. C. Olson, Ref. 11, p. 523.
18. P. G. Fothergill, *Nature*, Vol. 191, p. 425 (1961).
19. P. R. Ehrlich and R. W. Holm, *Science*, Vol. 137, p. 655 (1962).
20. R. Danson, *New Scientist*, Vol. 49, p. 35 (1971).
21. N. Macbeth, *Darwin Retried*, Gambit, Inc., Boston, 1971.
22. P. Grassé, *L'Évolution du Vivant*, Editions Albin Michel, Paris, 1973.
23. T. Dobzhansky, *Evolution*, Vol. 29, p. 376 (1975).
24. Creation Research Society, 2717 Cranbrook Road, Ann Arbor, Michigan, 48104.
25. D. M. S. Watson, *Nature*, Vol. 124, p. 233 (1929).
26. T. Dobzhansky, *Science*, Vol. 175, p. 49 (1972).
27. J. Huxley, *The Observer*, July 17, 1960, p. 17.
28. *What is Humanism?* A pamphlet published by The Humanist Community of San Jose, California, 95106.
29. G. G. Simpson, *Science*, Vol. 131, p. 966 (1960).
30. G. G. Simpson, *Life of the Past*, Yale University Press, New Haven, 1953.
31. J. Huxley, *Scientific American*, Vol. 189, p. 90 (1953).

Chapter II

THE CREATION AND EVOLUTION MODELS

Much evidence from the fields of cosmology, chemistry, thermodynamics, mathematics, molecular biology, and genetics could be inferred in an attempt to decide which model offers a more plausible explanation for the origin of living things. In the final analysis, however, what actually *did* happen can only be decided, scientifically, by an examination of the historical record, that is, the fossil record. Thus, W. LeGros Clark, the well-known British evolutionist, has said:

> That evolution actually *did* occur can only be scientifically established by the discovery of the fossilized remains of representative samples of those intermediate types which have been postulated on the basis of the indirect evidence. In other words, the really crucial evidence for evolution must be provided by the paleontologist whose business it is to study the evidence of the fossil record.[1]

This latter statement also applies to creation.

The history of life upon the earth may be traced through an examination of the fossilized remains of past forms of life entombed in the rocks. If life arose from an inanimate world through a mechanistic, naturalistic, evolutionary process and then diversified, by a similar process via increasingly complex forms, into the millions of species that have existed and now exist, then the fossils actually found in the rocks should correspond to those predicted on the basis of such a process.

On the other hand, if living things came into being by a process of special creation, then predictions very different from those based on evolution theory should be made concerning the fossil record. It is our contention that the fossil record is much more in accordance with the predictions based on creation, rather than those based on the theory of evolution, and actually strongly contradicts evolution theory. The purpose of this publication is to document that contention and to demonstrate that all of the facts derivable from the fossil record are readily correlated within a framework of special creation.

DEFINITIONS

For the purposes of this discussion, it is very important that it be clearly understood just what we mean by the terms evolution and creation.

Evolution. When we use the term *Evolution* we are using it in the sense defined by the general theory of evolution. According to the *General Theory of Evolution,* all living things have arisen by a naturalistic, mechanistic, evolutionary process from a single living source which itself arose by a similar process from a dead, inanimate world. This has sometimes been called the "amoeba-to-man" theory, or as I sometimes call it, the "fish-to-Gish" theory.

According to this theory, all living things are interrelated. Man and ape, for example, are believed to have shared a common ancestor. The divergence from this common ancestor has been variously estimated to have occurred from 5 to 30 million years ago, depending upon who is telling the story. The primates, which include men and apes, are believed to have shared a common ancestor with the horse, and this divergence is believed to have occurred approximately 75 million years ago.

Similar relationships are imagined throughout the entire animal and plant kingdoms. The supposed evolutionary relationships of an animal or plant with all other animals or plants is referred to as its phylogeny, and such relationshps are portrayed in a so-called phylogenetic tree. One such tree is illustrated in Figure 1.

Equally important to our discussion is an understanding of just what we are *not* talking about when we use the term evolution. We are not referring to the limited variations that can be seen to occur, or which can be inferred to have occurred in the past, but which do not give rise to a new basic kind.

We must here attempt to define what we mean by a basic kind. A basic animal or plant kind would include all animals or plants which were truly derived from a single stock. In present-day terms, it would be said that they have shared a common gene pool. All humans, for example, are within a single basic kind, *Homo sapiens*. In this case, the basic kind is a single species.

In other cases, the basic kind may be at the genus level. It may be, for instance, that the various species of the coyote, such as the Oklahoma Coyote

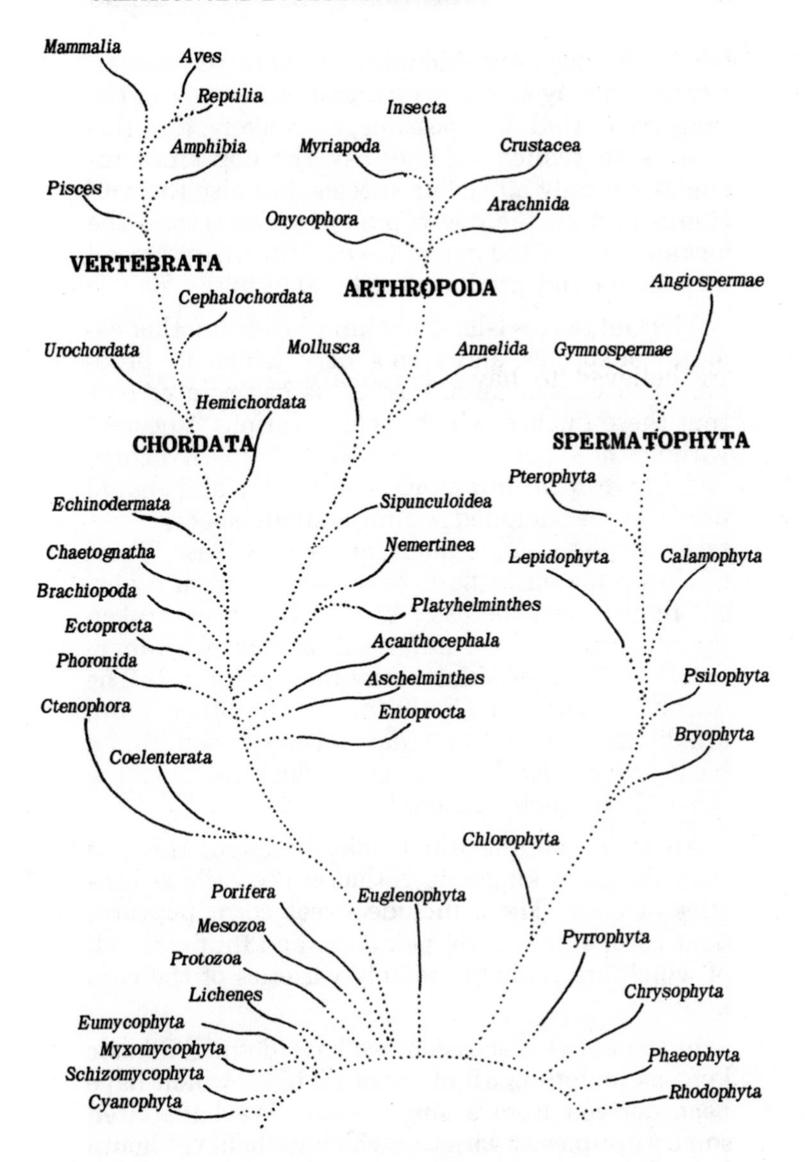

FIGURE 1. Hypothetical phylogenetic tree.

(*Canis frustor*), the Mountain Coyote (*C. lestes*), the Desert Coyote (*C. estor*), and others, are of the same basic kind. It is possible, even likely, that this basic kind (which we may call the dog kind) includes not only all coyote species, but also the wolf (*Canis lupus*), the dog (*Canis familiaris*), and the jackals, also of the genus *Canis*, since they are all interfertile and produce fertile offspring.

The Galapagos Island finches provide another example of species, and even genera, which are probably of one basic kind. Lammerts has pointed out[2] that these finches, which include various "species" within the "genera," *Geospiza, Camarhynchus,* and *Cactospiza*, intergrade completely and should probably be included within a single species, certainly within a single genus, at the very least. These finches apparently have been derived from a parent finch stock, the basic kind having been broken up into various forms as a result of chance arrangement of their original variability potential. The warbler finches, or *Certhidea*, on the other hand, are distinctive, and may have been derived from a basic stock separate from that which includes the other three finch "genera."

Another example which may be cited, this one from the plant kingdom, is that of the various varieties of corn. These include sweet corn, popcorn, dent corn, starch corn, pod corn, and flint corn, all of which are probably merely varieties of the corn kind.[3]

In the above discussion, we have defined a basic kind as including all of those variants which have been derived from a single stock. We have cited some examples of varieties which we believe should be included within a single basic kind. We cannot always be sure, however, what constitutes a separate kind. The division into kinds is easier the more the divergence observed. It is obvious, for example, that among invertebrates the protozoa, sponges, jellyfish, worms, snails, trilobites, lobsters, and bees are all different kinds. Among the vertebrates, the fishes, amphibians, reptiles, birds, and mammals are obviously different basic kinds.

Among the reptiles, the turtles, crocodiles, dinosaurs, pterosaurs (flying reptiles), and ichthyosaurs (aquatic reptiles) would be placed in different kinds. Each one of these major groups of reptiles could be further subdivided into the basic kinds within each.

Within the mammalian class, duckbilled platypuses, opposums, bats, hedgehogs, rats, rabbits, dogs, cats, lemurs, monkeys, apes, and men are easily assignable to different basic kinds. Among the apes, the gibbons, orangutans, chimpanzees, and gorillas would each be included in a different basic kind.

When we attempt to make fine divisions within groups of plants and animals where distinguishing features are subtle, there is a possibility of error. Many taxonomic distinctions established by man are uncertain and must remain tentative.

Let us now return to our discussion of evolution. According to the general theory of evolution, not only have the minor variations within kinds arisen through natural processes, but the basic kinds themselves have arisen from fundamentally different ancestral forms. Creationists do not deny the former, that is, the origin of variations within kinds, but they do deny the latter, that is, the evolutionary origin of basically different types of plants and animals from common ancestors.

In our discussion of evolution, therefore, we are *not* referring, for example, to the possible origin of the variations within the dog kind. We are referring to the alleged origin of the dog kind and cat kind from a common ancestor. We are *not* referring to the origin of the finches within *Geospiza, Camarhynchus,* and *Cactospiza*. We *are* referring to the origin of these finches and, say, the herons, from a common ancestor, and their ultimate origin from an ancestral reptile.

Neither are we referring to "industrial melanism,"[4] a case often cited by evolutionists as proof for evolution. The peppered moth, *Biston betularia,* is normally white with a covering of black spots and stripes. Melanic, or dark-colored, forms, known as the carbonaria form, have always existed, but as rarities.

Before the advent of the industrial revolution and resultant air pollution, the tree trunks in England were light-colored. The peppered moth rests on tree trunks during the day, with wings outspread. The normal, or light-colored, variant is very inconspicuous against such a background. The melanic form, on the other hand, is easily detected under these circumstances. As a result, predators (birds) picked off a much higher percentage of the melanic form, and they thus remained a minor proportion of the total population of peppered moths.

This was the case in 1850 at about the time the industrial revolution in England began. The tree trunks became progressively darker, however, and by 1895, 95% of these moths in the vicinity of Manchester were of the carbonaria, or melanic, variety. This change had taken place because now the melanic form was inconspicuous against the blackened tree trunks, while the light-colored variant was easily detected when resting against this background.

We wish to emphasize, first of all, that this process did not result in increasing complexity or even anything new. The melanic form of the peppered moth had existed in England many years before the industrial revolution. It was a stable, though minor, fraction of the population. The change brought about by air pollution did decrease detection of this preexisting form by its natural enemies and thus resulted in a shift in populations of the melanic versus the light-colored form.

Of greatest importance to our discussion, however, is the fact that no significant evolutionary change has occurred in these moths. These moths today not only are still moths, *but they are still peppered moths, Biston betularia.* This evidence, therefore, is irrelevant to the questions we seek to answer: did these lepidopterous insects arise by a naturalistic, mechanistic process from a nonlepidopterous insect? Did the insects themselves arise from a noninsect form of life?

While no real evolutionary change took place in the shift in populations of the two varieties of the peppered moth, one Natural Science encyclopedia has recently characterized this event as "the most striking evolutionary change ever to be witnessed by man."[5] If this is the best evidence for evolution that can be produced, then indeed—just as Dobzhansky has admitted—evolution has not been witnessed by human observers, for this is not evolution at all!

The evolutionist assumes that the accumulation of many such minor changes eventually could result in a new basic type and in increasing complexity, but *this is purely an assumption.* What is required is experimental evidence, or, lacking that, hard fossil evidence, or historical proof, that basic changes of this type actually did take place.

Another form of change which is often cited by evolutionists as evidence for evolution is the origin of domesticated plants and animals by artificial selection and breeding. Evidence of this nature is again irrelevant to our discussion, since nothing new or more complex arises, and the change accomplished is always extremely limited.

What artificial selection and breeding actually accomplishes is to rapidly establish the *limit* beyond which no further change is possible. We wish to cite just two examples.[6] In 1800, experiments were begun in France to increase the sugar content of table beets, which at that time amounted to 6%. By 1878, the sugar content had been increased to 17%. Further selection failed to increase the sugar content above that figure.

One worker tried to reduce the number of bristles on the thorax of fruit flies by artificial selection and breeding. In each generation, the average number of bristles became fewer until the twentieth generation. After that, the average remained the same, although he selected as before. Selection was no longer effective; the limit had been reached.

Similar experimental approaches have been used to develop chickens that lay more eggs, cows that give more milk, and corn with increased protein content. In each case, limits were reached beyond which further change has not been possible. Furthermore, the breeders ended up with the same species of chickens, cows, and corn with which they began. No real change had taken place.

It must be strongly emphasized, also, that in all

cases these specialized breeds possess reduced viability; that is, their basic ability to survive has been weakened. Domesticated plants and animals do not compete well with the original, or wild type. Thus, Falconer has stated:

> Our domesticated animals and plants are perhaps the best demonstration of the effects of this principle. The improvements that have been made by selection in these have clearly been accompanied by a reduction of fitness for life under natural conditions, and only the fact that domesticated animals and plants do not live under natural conditions has allowed these improvements to be made.[7]

These experiments thus demonstrate that even with the aid of man's inventive genius, which permits the maximum variation in the shortest possible time, the variation achieved is extremely limited and actually results in plants and animals with reduced viability. They survive only because they are maintained in an environment which is free from their natural enemies, food supplies are abundant, and other conditions are carefully regulated.

In summary, then, by evolution we mean a process which is supposed to have been responsible for converting the most primitive form of life, the hypothetical primordial cell, via innumerable increasingly complex forms of life, into man, the highest form of life. The theory of evolution, then, proposes that basically different types of plants and animals have arisen from common ancestors, which in turn had arisen from more ancient and more primitive forms of life. By evolution, we do not mean limited variations that have taken place within a distinct, separate kind, and which have not led to the origin of a basically different form of life.

Creation. By creation we mean the bringing into being by a supernatural Creator of the basic kinds of plants and animals by the process of sudden, or fiat, creation.

We do not know how the Creator created, what processes He used, *for He used processes which are not now operating anywhere in the natural universe.* This is why we refer to creation as special creation. We cannot discover by scientific investigations anything about the creative processes used by the Creator.

In our earlier discussion, we have defined what we mean by a basic animal or plant kind. During creation the Creator created all of these basic animal and plant kinds, and since then no new kinds have come into being. The variation that has occurred since the end of creation has been limited to changes within kinds.

As noted earlier, then, the concept of special creation does not exclude the origin of varieties and species from an original created kind. It is believed that each kind was created with sufficient genetic potential, or gene pool, to give rise to all of the varieties within that kind that have existed in the past and those that are yet in existence today.

Each kind was created with a great variety of genes. These genes can be sorted out during the sexual reproductive process in an enormous number of different ways. For instance, there are approximately 4 billion human beings in the world today, and except for identical twins and other cases of multiple births, no two individuals are exactly alike. None have the same gene combination. This sorting out process has not only given rise to many different individuals but also to distinctively different races. All remain, however, members of one species, *Homo sapiens*.

Another example familiar to all of us is that of the dog. All of the dogs, from the tiny Chihuahua to the Great Dane, and from the bulldog to the greyhound have been derived from a single species, *Canis familiaris*. The process has been exaggerated by man, of course, through artificial selection and inbreeding.

Many other examples could be cited. In each case the great variety of genes responsible for the variations that have taken place was present in the original created kind. There has merely occurred a sorting out in many different ways. No matter what combinatons may occur, however, the human kind always remains human, and the dog kind never ceases to be dog kind. The transformations proposed by the theory of evolution never take place.

The Evolutionary Mechanism. Before we can evaluate the fossil record for the evidence it can shed on the question of creation and evolution, we must first understand the mechanism by which evolution has supposedly occurred. On the basis of this hypothetical mechanism, we will be able to predict what the fossil record ought to show if evolution actually did occur.

We have noted above the many variations that exist within each kind. Darwin had noted this fact, although he did not understand what was responsible for the origin of this variability. Darwin proposed that changes are constantly taking place within a species.

Darwin was aware of the fact that many more

animals are born than actually survive. He envisioned a struggle for existence in which the stronger survive and the weaker are eliminated. Under these conditions any variation that results in a lowered viability (basic ability to survive) or reproductive capacity would cause the elimination of the plant or animal inheriting this variation.

On the other hand, Darwin reasoned, any variation which increased the viability or fertility of a plant or animal would give it an advantage in the struggle for existence. This favored variant and its offspring which inherited this favorable variation would tend to survive at the expense of the unchanged variety. Nature was said to have selected the favored variant, and the evolutionary process was thus said to consist of *variation with natural selection.* The accumulation of many of these small hypothetically favorable changes over a long period of time supposedly was able to accomplish the most profound changes, even converting a microscopic bacterial cell into a human being.

Darwin knew nothing about what was responsible for the variability within species. Gregor Mendel's great work on genetics was published at about the same time as Darwin's *Origin of Species*, but was ignored by Darwin and all other scientists at that time. What Darwin did propose to account for the origin of variability was completely erroneous. He accepted the idea of the inheritance of acquired characteristics. This is the idea that when cells in the tissues (somatic cells) are affected by the environment, hereditary units ("gemmules") are formed. These "gemmules," it was believed, were carried to the germ cells and then passed on to the offspring. The characteristic acquired by the parent was thus supposedly inherited by their offspring.

Today we know that inheritance is controlled by the genes found solely in the germ cells (the eggs, or ova, and the spermatazoa). Only alterations in the genes of the germ cells are inheritable. No such thing as a "gemmule" is formed, and acquired characteristics are not inherited.

Hundreds of thousands of genes are present in the nucleus of every cell of the higher animals. Each gene consists of a long strand of several hundred to several thousand subunits, linked together like the links of a chain. The particular type of complex chemical which constitutes a gene is called deoxyribonucleic acid, or DNA.

There are four different kinds of subunits (nucleotides) in DNA. It is the particular order of these subunits in the DNA chain which distinguishes one gene from another, just as the specific sequence of the letters of the alphabet distinguishes one sentence from another.

Each characteristic is influenced by at least two genes. The genes of this gene pair are called alleles. One such gene is inherited from each of the parents. Thus, the egg and sperm each have a single set of genes. When fertilization occurs, these two sets of genes combine. The segregation and recombination of the genes which occur during production of the germ cells produce sperm and eggs with a tremendous variety of different gene combinations. These sperm and egg cells in turn, via random mating, can be combined in a great variety of ways. The result is the tremendous variability that we see within each species.

The genes are ordinarily very stable. A particular gene (in the form of its successors) may exist many thousands of years without alteration in its structure. Very rarely, however, the chemical structure of a gene does undergo a change. Such a change is called a mutation. Mutations may be caused by chemicals, X-rays, ultraviolet light, cosmic rays, and other causes. Some may occur during cell reproduction due to copying errors.

Most mutations result in a change in only one of the several hundred or several thousand subunits in a gene. The change usually is so subtle that it cannot be directly detected by present chemical techniques. The effect on the plant or animal almost always is very drastic, however. Frequently, a mutation proves to be lethal, and it is almost universally, or is universally, harmful.

The mutations we see occurring spontaneously in nature or that can be induced in the laboratory always prove to be harmful. It is doubtful that of all the mutations that have been seen to occur, a single one can definitely be said to have increased the viability of the affected plant or animal.[8] Evolutionists claim, however, that a very small fraction (perhaps 1 in 10,000) of these mutations are beneficial. This claim is made, not because we can actually observe such favorable mutations occurring, but because evolutionists know that unless favorable mutations do occur, evolution is impossible. In the final analysis, all of evolution must be ascribed to mutations.[9]

These hypothetical beneficial mutations supposedly alter the plant or animal in such a way that its ability to compete and survive is enhanced and/or its reproductive capacity is increased. The plants or animals inheriting these mutant genes then would tend to survive at the expense of the unchanged variety. Evolutionists believe that after many thousands of generations eventually the mutant would completely replace the original, unchanged variety. Nature is said to have selected the favored mutant, and the evolutionary process, therefore, is termed *mutation with natural selection.*

Evolutionists, with very few exceptions, believe that these proposed favorable mutations which supposedly contribute to evolution must result in only slight changes. A mutation that would result in more than a slight change would be too disruptive to the organization of the plant or animal for it to survive. Such a mutation would be certain to be lethal.

Since each one of the mutations which have supposedly contributed to the evolutionary process would have resulted in only a very slight change, it is evident that the evolution of one species into another would require the accumulation of many thousands of these hypothetical favorable mutations. A much more drastic change, such as the conversion of a fish into an amphibian, would require a very large number of favorable mutations in many, many characteristics.

A mutation of any kind in a gene is a rare event. Furthermore, if only 1 out of 10,000 or less of these mutations is favorable, it can be seen that the occurrence of a favorable mutation is indeed an extremely rare event, assuming that they occur at all. Furthermore, in order to be inheritable, a mutation must occur in the genes of the germ cells. The germ cells make up only a tiny fraction of all the cells of an organism and are generally relatively well-protected from the environment. It is obvious that *the essence of the postulated evolutionary process is slow and gradual change.* The change of a species into a new species is believed to require hundreds of thousands, if not millions, of years. A drastic change such as the change of fish into amphibian or reptile into mammal is believed to have required several tens of millions of years.

The interpretation of the evolutionary process as very slow and gradual change due to small, or micromutations, in combination with the sorting out that accompanies sexual reproduction, all influenced by natural selection in accordance with the environment, is termed the neo-Darwinian interpretation of evolution. The essence of Darwinism is retained but Darwin's theories have been modified to accommodate them to the discoveries made since his time in genetics, molecular biology, etc. With very few exceptions, all evolutionists are neo-Darwinists.

PREDICTIONS BASED ON THE CREATION MODEL AND THE EVOLUTION MODEL

In the preceding discussion we have defined what is meant by creation and evolution. We have described the evolutionary mechanism that is accepted by almost all evolutionists, and we have described all that human knowledge will permit about the creation process. We are now ready to predict the evidence that must be found in the fossil record based on the creation model on the one hand and the evolution model on the other hand.

Creation Model. On the basis of the creation model, we would predict an explosive appearance in the fossil record of highly complex forms of life without evidence of ancestral forms. We would predict that all of the major types of life, that is, the basic plant and animal forms, would appear abruptly in the fossil record without evidence of transitional forms linking one basic kind to another.

We would thus expect to find the fossilized remains, for example, of cats, dogs, bears, elephants, cows, horses, bats, dinosaurs, crocodiles, monkeys, apes, and men without evidence of common ancestors. Each major kind at its earliest appearance in the fossil record would possess, fully developed, all the characteristics that are used to define that particular kind.

Evolution Model. On the basis of the evolution model, we would predict that the most ancient strata in which fossils are found would contain the most primitive forms of life capable of leaving a fossil record. As successively younger strata were searched, we would expect to see the gradual transition of these relatively simple forms of life into more and more complex forms of life. As living forms diverged into the millions of species which have existed in the past and which exist today, we would expect to find a slow and gradual transition of one form into another.

We would predict that new basic types would *not* appear suddenly in the fossil record possessing all of the characteristics that are used to define its kind. The earliest forms in each group would be expected to possess in incipient form some of the characteristics which are used to define that group while retaining characteristics used to define the ancestral group.

If fish evolved into amphibia, as evolutionists believe, then we would predict that we would find transitional forms showing the gradual transition of fins into feet and legs. Of course, many other alterations in the anatomy and physiology of fishes would have to occur to change an animal adapted to living its entire life span in water to one which spends most of its life outside of water. The fin-to-feet transition would be an easily traceable transition, however.

If reptiles gave rise to birds, then we would expect to find transitional forms in the fossil record showing the gradual transition of the forelimbs of the ancestral reptile into the wings of a bird, and the gradual transition of some structure on the reptile into the feathers of a bird. These again are obvious transitions that could be easily traced in the fossil record. Of course, many other changes would have been taking place at the same time, such as the conversion of the hindfeet of the reptile into the perching feet of the bird, reptilian skull into bird-like skull, etc.

In the pterosaurs, or flying reptiles, the wing membrane was supported by an enormously lengthened fourth finger. If the pterosaurs actually evolved from a nonflying reptile, then we would predict that the fossil record would produce transitional forms showing a gradual increase in length of the fourth finger, along with the origin of other unique structures.

The fossil record ought to produce thousands upon thousands of transitional, or in-between forms. It is true that according to evolutionary geology only a tiny fraction of all plants and animals that have ever existed would have been preserved as fossils. It is also true that we have as yet uncovered only a small fraction of the fossils that are entombed in the rocks. We have, nevertheless, recovered a good representative number of the fossils that exist.

Sampling of the fossil record has now been so thorough that appeals to the imperfections in the record are no longer valid. George has stated:

> There is no need to apologize any longer for the poverty of the fossil record. In some ways it has become almost unmanageably rich and discovery is outpacing integration.[10]

It seems clear, then, that after 150 years of intense searching a large number of obvious transitional forms would have been discovered if the predictions of evolution theory are valid.

We have, for example, discovered literally billions of fossils of ancient invertebrates and many fossils of ancient fishes. The transition of invertebrate into vertebrate is believed to have required many millions of years. Populations are supposed to constitute the units of evolution and, of course, only successful populations survive. It seems obvious, then, that if we find fossils of the invertebrates which were supposed to have been ancestral to fishes, and if we find fossils of the fishes, we surely ought to find the fossils of the transitional forms.

We find fossils of crossopterygian fishes which are alleged to have given rise to the amphibia. We find fossils of the so-called "primitive" amphibia. Since the transition from fish to amphibia would have required many millions of years, during which many hundreds of millions, even billions, of the transitional forms must have lived and died, many of these transitional forms should have been discovered in the fossil record even though only a minute fraction of these animals have been recovered as fossils. As a matter of fact, the discovery of only five or six of the transitional forms scattered through time would be sufficient to document evolution.

So it would be throughout the entire fossil record.

There should not be the slightest difficulty in finding transitional forms. Hundreds of transitional forms should fill museum collections. If we find fossils at all, we ought to find transitional forms. As a matter of fact, difficulty in placing a fossil within a distinct category should be the rule rather than the exception.

SUMMARY

The contrast between the two models and the predictions based on each model may be summarized as follows:

Creation Model	**Evolution Model**
By acts of a Creator.	By naturalistic mechanistic processes due to properties inherent in inanimate matter.
Creation of basic plant and animal kinds with ordinal characteristics complete in first representatives.	Origin of all living things from a single living source which itself arose from inanimate matter. Origin of each kind from an ancestral form by slow gradual change.
Variation and speciation limited within each kind.	Unlimited variation. All forms genetically related.

These two models would permit the following predictions to be made concerning the fossil record:

Creation Model	Evolution Model
Sudden appearance in great variety of highly complex forms.	Gradual change of simplest forms into more and more complex forms.
Sudden appearance of each created kind with ordinal characteristics complete. Sharp boundaries separating major taxonomic groups. No transitional forms between higher categories.	Transitional series linking all categories. No systematic gaps.

REFERENCES

1. W. LeGros Clark, *Discovery*, p. 7, January, 1955.
2. W. E. Lammerts, "The Galapagos Island Finches," in *Why Not Creation?* W. E. Lammerts, ed., Presbyterian & Reformed Publ. Co., Philadelphia, 1970, p. 354.
3. F. L. Marsh, *Creation Research Society Quarterly*, Vol. 8, p. 13 (1969).
4. W. Wickler, *Mimicry in Plants and Animals.* World University Library, New York, 1968, p. 51.
5. *The International Wildlife Encyclopedia*, M. Burton and R. Burton, eds., Marshal Cavendish Corp., New York, 1970, p. 2706.
6. W. J. Tinkle, *Heredity*, St. Thomas Press, Houston, 1967, p. 55.
7. D. S. Falconer, *Introduction to Quanitative Genetics*, Ronald Press, 1960, p. 186.
8. C. P. Martin, *American Scientist*, Vol. 41, p. 100 (1953).
9. E. Mayr, in *Mathematical Challenges to the Neo-Darwinian Interpretation of Evolution*, P. S. Moorhead and M. M. Kaplan, Eds., Wistar Institute Press, Philadelphia, 1967, p. 50.
10. T. N. George, *Science Progress*, Vol. 48, p. 1 (1960).

Chapter III

GEOLOGIC TIME AND THE GEOLOGIC COLUMN

With few exceptions, such as the La Brea Tar Pits, fossils are found in sedimentary deposits. The formation of sedimentary rocks involves erosion, transportation, deposition, and lithification. The action of wind, freezing and thawing, rain, and flooding have caused rocks to disintegrate. The resultant particles, ranging in size from extremely fine particles to boulders, have been transported by water (some have been transported by wind, glaciers, and other agencies, but these represent exceptional cases) and then deposited when the water reached a quiet area. Through the action of cementing agents and/or pressure, these deposits have become consolidated in the form of sedimentary rocks.

The hard parts of marine organisms may be preserved in marine sediments. Fresh water organisms, land animals, and plants may be entrapped and swept along by moving water and buried with the sediments. As the sediments become compacted into rocks, the bones of animals or the imprint left by the remains of animals and plants may become part of the rocks. These remains are known as fossils. Some sedimentary deposits are a few feet thick, some are hundreds of feet thick, and rarely, some are even a thousand feet or more in thickness.

Several approaches to the interpretation of geologic history have been applied.

UNIFORMITARIANISM.

The uniformitarian concept of historical geology is accepted by almost all evolutionists. According to this interpretation of earth history, existing physical processes, acting essentially at present rates, are sufficient to account for all geological formations. As originally formulated by James Hutton and Charles Lyell, any appeal to catastrophes for the explanation of geologic phenomena are rejected. The phrase "the present is the key to the past" was coined for this concept.

According to this interpretation, the formation of sedimentary deposits hundreds of feet thick would have required millions of years. It was also realized that evolution would have required many millions of years. Accordingly, the age of the earth as estimated by evolutionary geologists began to increase at an astounding rate. The application of certain assumptions with radiometric dating methods finally has allowed present-day geologists to estimate an age of about 4.5 billion years for the earth.

Geologists have classified sedimentary deposits according to the type of fossils found in the deposits. Certain fossils are believed to have been laid down during a restricted time span. These fossils have been designated as "index fossils" and are used by evolutionists to identify and date rocks. For example, any rock containing fossils of a certain type of trilobite is designated as a Cambrian rock.

Evolutionists assume that Cambrian sedimentary rocks were deposited during a stretch of approximately 80 million years beginning about 600 million years ago. This period has been named the Cambrian Period. They assume that other sedimentary deposits have followed in a chronological order, each spanning millions of years. The Cambrian Period is assumed to have been followed by the Ordovician, Silurian, Devonian, Mississippian, etc.

This arrangement of various types of fossiliferous deposits in a supposed time-sequence is known as the *geological column.* Its arrangement is based on the assumption of evolution. Thus, invertebrates are assumed to have evolved first, followed by fish, amphibia, reptiles, and mammals in that order. The geological column has been arranged accordingly.

The above brief description of the uniformitarian concept of historical geology is necessarily sketchy and simplified. Any standard textbook on geology may be consulted for a more thorough description of this system.

MODIFIED UNIFORMITARIAN CONCEPTS

The Day-Age Theory. Some creationists accommodate the uniformitarian concept of historical geology by assuming that creation occurred sporadically through long periods of time. It is assumed that the Creator allowed varying periods of time to intervene between successive creations, and that animals and plants were created in the sequence required by the assumed geological column.

GEOLOGIC TIME TABLE

MAIN DIVISIONS OF GEOLOGICAL TIME

ERAS	PERIODS	ESTIMATED YEARS AGO
CENOZOIC	Quaternary:	
	Recent Epoch	25,000
	Pleistocene Epoch	3,000,000
	Tertiary:	
	Pliocene Epoch	12,000,000
	Miocene Epoch	25,000,000
	Oligocene Epoch	35,000,000
	Eocene Epoch	60,000,000
	Paleocene Epoch	70,000,000
MESOZOIC	Cretaceous Jurassic Triassic	70,000,000 to 200,000,000
PALEOZOIC	Permian Pennsylvanian Mississippian Devonian Silurian Ordovician Cambrian	200,000,000 to 600,000,000
PROTEROZOIC		600,000,000 to 1,000,000,000
ARCHEOZOIC		1,000,000,000 to 1,800,000,000

The Gap Theory. According to this theory, there was an initial creation spanning geological ages. A great time span then followed. The geological column is believed to have formed during this initial period of creation and subsequent time span.

Then the Creator is said to have destroyed His original creation for some reason. A second creation in six literal 24-hour days is then believed to have taken place. The gap theory is an attempt to accommodate both the geological column, with its vast time span, and a six 24-hour day creation.

THE CATASTROPHIST — RECENT CREATION MODEL.

The proponents of this model for interpreting geological history believe that creation spanned six 24-hour days. Furthermore, it is believed creation occurred thousands rather than billions of years ago.

While present geological processes may have operated at present rates for long periods of time, the advocates of this model contend that it is impossible to account for most of the important geological formations according to uniformitarian principles. These formations include the vast Tibetan Plateau, 750,000 square miles of sedimentary deposits many thousands of feet in thickness and now at an elevation of three miles; the Karoo Formation of Africa, which has been estimated by Robert Broom to contain the fossils of 800 billion vertebrate animals[1]; the herring fossil bed in the Miocene shales of California, containing evidence that a billion fish died within a four-square mile area[2]; and the Cumberland Bone Cave of

Maryland, containing fossilized remains of dozens of species of mammals, from bats to mastodons, along with the fossils of some reptiles and birds — including animals which now have accommodated to different climates and habitats from the Arctic region to tropical zones.[3] Neither has the uniformitarian concept been sufficient to explain mountain building nor the formation of such vast lava beds as the Columbian Plateau in northwest United States, a lava bed several thousand feet thick covering 200,000 square miles.

It is believed that most of the important geological formations of the earth can be explained as having been formed as the result of a worldwide flood, along with attendant vast earth movements, volcanic action, dramatic changes in climatic conditions, and other catastrophic events. The fossil record, rather than being a record of transformation, is a record of mass destruction, death, and burial by water and its contained sediments. Proponents of this interpretation of earth history not only face the unenviable position of being labeled as rank heretics, but a massive reexamination and reinterpretation of geologic data is required. The reinterpretation of geologic data according to flood geology would include an evaluation of all dating methods, including especially a critical review of radiometric dating methods. Such work is already well under way. It should be realized that there is no *direct* method for determining the age of any rock. While very accurate methods are available for determining the *present* ratios of uranium-lead, thorium-lead, potassium-argon, and other isotope ratios in mineral-bearing rocks, there is, of course, no direct method for estimating the *initial* ratios of these isotopes in the rocks when the rocks were first formed. Radiochronologists must resort to indirect methods which involve certain basic assumptions. Not only is there no way to verify the validity of these assumptions, but inherent in these assumptions are factors that assure that the ages so derived, whether accurate or not, will always range in the millions to billions of years (excluding the carbon-14 method, which is useful for dating samples only a few thousand years old).

Recent publications[4-8] have exposed weaknesses in radiometric dating methods, while some recent publications[4-5, 8-12] have described many reliable chronometers, or "time-clocks," that indicate a young age for the earth. Discussions of the catastrophist interpretation of historical geology may be found in a number of books and recent publications.[4-5, 13-21]

In order to evaluate evolution as an interpretive model to explain origins, and to compare the predictive value of this model to that of the creation model, the assumptions of evolutionary geologists concerning the duration of geological ages and the validity of their assumptions concerning the geological column must be used along with the model. Therefore, in the succeeding pages of this book we will write as though the Cambrian, Ordovician, Silurian, and other sedimentary deposits were actually laid down during the time spans generally assumed by evolutionists, and that the arrangement of the geological column in the form of successive geological periods as accepted by evolutionary geologists is correct.

Even when these assumptions are accepted, however, the data from the fossil record does not agree with the predictions of the evolution model.

Therefore, whether or not the earth is ten thousand, ten million, or ten billion years old, the fossil record does not support the general theory of evolution.

REFERENCES

1. N. O. Newell, *Journal of Paleontology*, Vol. 33, p. 496 (1959).
2. H. S. Ladd, *Science*, Vol. 129, p. 72 (1959).
3. G. Nicholas, *Scientific Monthly*, Vol. 76, p. 301 (1953).
4. J. C. Whitcomb and H. M. Morris, *The Genesis Flood*, Presbyterian and Reformed Publ. Co., Philadelphia, 1964.
5. M. A. Cook, *Prehistory and Earth Models*, Max Parrish and Co., Ltd., London, 1966.
6. H. Slusher, *Critique of Radiometric Dating Methods*, Creation-Life Publishers, Inc., San Diego, 1973.
7. S. P. Clementson, *Creation Research Society Quarterly*, Vol. 7, p. 137 (1970).
8. M. A. Cook, *Creation Research Society Quarterly*, Vol. 7, p. 53 (1970).
9. R. L. Whitelaw, in *Why Not Creation?* W. E. Lammerts, ed., Presbyterian and Reformed Publ. Co., Philadelphia, 1970, pp. 90 and 101.
10. R. V. Gentry, in Ref. 9, p. 106.
11. H. S. Slusher, *Creation Research Society Quarterly*, Vol. 8, p. 55 (1971).

12. T. G. Barnes, *Origin and Destiny of the Earth's Magnetic Field*, Creation-Life Publishers, Inc., San Diego, 1973.
13. G. M. Price, *Evolutionary Geology and the New Catastrophism*, Pacific Press Pub. Assoc., Mountain View, California, 1926.
14. H. M. Morris, *Biblical Cosmology and Modern Science*, Baker Book House, Grand Rapids, Mich., 1970.
15. H. W. Clark, *Fossils, Flood, and Fire*, Outdoor Pictures, Escondido, California, 1968.
16. H. M. Morris, in Ref. 9, p. 114.
17. H. M. Morris, in *Scientific Studies in Special Creation*, W. E. Lammerts, ed., Presbyterian and Reformed Publ. Co., Philadelphia, 1971, p. 103.
18. N. A. Rupke, in Ref. 9, p. 141.
19. C. L. Burdick, in Ref. 17, p. 125.
20. H. W. Clark, in Ref. 17, p. 156.
21. E. C. Powell, *Creation Research Society Quarterly*, Vol. 9, p. 230 (1973).

Chapter IV

THE FOSSIL RECORD — FROM MICROORGANISMS TO FISHES

LIFE APPEARS ABRUPTLY IN HIGHLY DIVERSE FORMS

The oldest rocks in which indisputable metazoan fossils are found are those of the Cambrian Period. In these sedimentary deposits are found billions and billions of fossils of highly complex forms of life. These include sponges, corals, jellyfish, worms, mollusks, crustaceans; in fact, every one of the major invertebrate forms of life have been found in Cambrian rocks. These animals were so highly complex it is conservatively estimated that they would have required 1½ billion years to evolve.

What do we find in rocks older than the Cambrian? *Not a single, indisputable, metazoan fossil has ever been found in Precambrian rocks!* Certainly it can be said without fear of contradiction that the evolutionary ancestors of the Cambrian fauna, if they ever existed, have never been found.[1-3]

Despite claims to the contrary, reports concerning the discovery of microfossils (microscopic single-celled bacteria and algae) in Precambrian rocks, estimated to be one to two billion years older than Cambrian, are questionable and certainly open to dispute. Some recent reports have tended to expose the uncertainties involved in such identifications.[4-7] For example, although they accepted the probability that certain alleged microfossils of Precambrian age were of biological origin, Engel, et al, cautioned that:

> Establishing the presence of biological activity during the very early Precambrian clearly poses difficult problems. . . . skepticism about this sort of evidence of early Precambrian life is appropriate.[7]

Even if these alleged microfossils represent the remains of genuine Precambrian forms of life, however, we are left with a tremendous gap between single-celled microscopic forms of life and the highly complex, highly diverse forms found in the Cambrian, a gap supposedly spanning one to two billion years of geologic time.

As recently as 1973, Preston Cloud, an evolutionary geologist, expressed his conviction that there are as yet no records of unequivocal Metazoa (multicellular forms of life) in undoubted Precambrian rocks.[2] Concerning this problem, Axelrod has stated:

> One of the major unsolved problems of geology and evolution is the occurrence of diversified, multicellular marine invertebrates in Lower Cambrian rocks on all the

> continents and *their absence in rocks of greater age.* (Emphasis added.)

After discussing the varied types that are found in the Cambrian, Axelrod goes on to say:

> However, when we turn to examine the Precambrian rocks for the forerunners of these Early Cambrian fossils, *they are nowhere to be found.* Many thick (over 5,000 feet) sections of sedimentary rock are now known to lie in unbroken succession below strata containing the earliest Cambrian fossils. These sediments apparently were suitable for the preservation of fossils because they are often identical with overlying rocks which are fossiliferous, *yet no fossils are found in them.*[3] (Emphasis added.)

From all appearances, then, based on the known facts of the historical record, there occurred a sudden great outburst of life at a high level of complexity. The fossil record gives no evidence that these Cambrian animals were derived from preceding ancestral forms. Furthermore, not a single fossil has been found that can be considered to be a transitional form between the major groups, or phyla. At their earliest appearance, these major invertebrate types were just as clearly and distinctly set apart as they are today. Thus, trilobites have always been trilobites, brachiopods have always been brachiopods, corals have always been corals, jellyfish have always been jellyfish, etc., and the fossil record offers no indication whatsoever that these highly varied invertebrate types have been derived from common ancestors.

How do these facts compare with the predictions

of the evolution model? *They are in clear contradiction to such predictions.* This has been admitted, for instance, by George who stated:

> Granted an evolutionary origin of the main groups of animals, and not on an act of special creation, the absence of any record whatsoever of a single member of any of the phyla in the Precambrian rocks remains as inexplicable on orthodox grounds as it was to Darwin.[8]

Simpson has struggled valiantly but not fruitfully with this problem, being forced to concede that the absence of Precambrian fossils (other than alleged fossil microorganisms) is the "major mystery of the history of life."[9]

These facts, however, are in full agreement with the predictions of the creation model. The fossil record *does* reveal a sudden appearance in great variety of highly complex forms with no evolutionary ancestors and *does* show the absence of transitional forms between the major taxonomic groups, just as postulated on the basis of creation. Thus, in a most emphatic manner, the known facts of the fossil record from the very outset support the predictions of the creation model, but unquestionably contradict the predictions of the evolution model.

The remainder of the history of life reveals a remarkable absence of the many transitional forms demanded by the theory. There is, in fact, a *systematic* deficiency of transitional forms between the higher categories, just as predicted by the creation model.

THE GREAT GULF BETWEEN INVERTEBRATE AND VERTEBRATE

The idea that the vertebrates were derived from the invertebrates is purely an assumption that cannot be documented from the fossil record. On the basis of comparative anatomy and embryology of living forms, almost every invertebrate group has been proposed at one time or another as the ancestor of the vertebrates.[10, 11] The transition from invertebrate to vertebrate supposedly passed through a simple, chordate stage. Does the fossil record provide evidence for such a transition? Not at all.

Ommaney has thus stated:

> How this earliest chordate stock evolved, what stages of development it went through to eventually give rise to truly fish-like creatures, we do not know. Between the Cambrian, when it probably originated, and the Ordovician, when the first fossils of animals with really fish-like characteristics appeared, there is a gap of perhaps 100 million years which we will probably never be able to fill.[12]

Incredible! One hundred million years of evolution and no fossilized transitional forms! All hypotheses combined, no matter how ingenious, could never pretend, on the basis of evolution theory, to account for a gap of such magnitude. Such facts, on the other hand, are in perfect accord with the predictions of the creation model.

DISTINCT SEPARATION OF MAJOR FISH CLASSES

A careful reading of Romer's book, *Vertebrate*

Paleontology,[11] seems to allow no other conclusions but that all of the major fish classes are clearly and distinctly set apart from one another with no transitional forms linking them. The first to appear in the fossil record is the class Agnatha. The most ancient of these vertebrates, representatives of the two orders Osteostraci and Heterostraci, were almost always encased in bony or other hard material and most were equipped with bony armor plates. Concerning their origin, Romer has written:

> In sediments of late Silurian and early Devonian age, numerous fishlike vertebrates of varied type are present, and it is obvious that a long evolutionary history had taken place before that time. But of that history we are mainly ignorant (p. 15).

Concerning the ostracoderms (Osteostraci) he writes, "When we first see these ostracoderms, they already have a long history behind them and are divided into several distinct groups" (p. 16). Of the Heterostraci, Romer writes that they are obviously not closely related to the other forms in the class Agnatha. If they evolved, they also must have had a long evolutionary history. But, like the ostracoderms, they appear suddenly in the fossil record without any evidence of evolutionary ancestors.

The placoderms are especially troublesome. Within the placoderms were about six major different kinds of weird fishes. Of them, Romer says, "There are few common features uniting these groups other than the fact that they are, without exception, peculiar" (p. 24). Later he says:

> They appear at a time — at about the Silurian-Devonian boundary — when we

> would expect the appearance of proper ancestors for the sharks and higher bony fish groups. We would expect "generalized" forms that would fit neatly into our preconceived evolutionary picture. Do we get them in the placoderms? Not at all. Instead, we find a series of wildly impossible types which do not fit into any proper pattern; which do not, at first sight, seem to come from any source or to be appropriate ancestors to any later or more advanced types. In fact, one tends to feel that the presence of these placoderms, making up such an important part of the Devonian fish story, is an incongruous episode; *it would have simplified the situation if they had never existed!* (p. 33). (Emphasis added.)

But they did exist, and their record does not support, but rather strongly contradicts, the evolution model.

The higher "orthodox" fish types, structured along well-recognized plans, with paired fins and well-developed jaws, are placed within two classes, the Chondrichthyes, or cartilaginous fishes, and the Osteichthyes, or higher bony fishes.

Some have argued in the past that the absence of bone in the cartilaginous fishes represents a primitive condition, and that the Chondricthyes were an evolutionary stage preceding bony fishes. Romer argues strongly against this, pointing out that the sharks were one of the last of the major fish groups to appear in the fossil record. He goes on to say:

> The record, in fact, fits in better with the opposite assumption: that the sharks

> are degenerate rather than primitive in their skeletal characters; that their evolution has paralleled that of various other fish types in a trend toward bone reduction; and that their ancestry is to be sought among primitive bony, jaw-bearing fishes of the general placoderm type. *No well-known placoderms can be identified as the actual ancestors of the Chondrichthyes*, but we have noted that some of the peculiar petalichthyids appear to show morphologically intermediate stages in skeletal reduction. Increasing knowledge of early Devonian placoderms may some day bridge the gap (p. 38). (Emphasis added.)

Earlier, with reference to the placoderms, Romer had said, ". . . we must consider seriously that at least the sharks and chimeras may have descended from such impossible ancestors" (p. 34). Romer insists that special creation is not admissible as a scientific explanation for origins, but he is willing to appeal to "impossible ancestors" to support his own sagging theory! A consideration of the creation model certainly seems more reasonable than an appeal to "impossible ancestors."

Of the typical bony fishes, Romer records the fact that their appearance in the fossil record is a "dramatically sudden one" (p. 52). Later on (p. 53), he states:

> *The common ancestor of the bony-fish groups is unknown.* There are various features, many of them noted above, in which the two typical subclasses of bony fish *are already widely divergent* when we first see them. . . . (Emphasis added.)

Errol White, an evolutionist and expert on fishes, in his presidential address on lungfishes to the Linnaean Society of London, said:

> But whatever ideas authorities may have on the subject, the lungfishes, like every other major group of fishes that I know, have their origins firmly based in *nothing*. . . .[13]

Later he went on to say, "I have often thought how little I should like to have to prove organic evolution in a court of law." He closed out his address by stating:

> We still do not know the mechanics of evolution in spite of the over-confident claims in some quarters, nor are we likely to make further progress in this by the classical methods of paleontology or biology; and we shall certainly not advance matters by jumping up and down shrilling "Darwin is God and I, So-and-so, am his prophet" — the recent researches of workers like Dean and Henshelwood (1964) already suggest the possibility of incipient cracks in the seemingly monolithic walls of the Neo-Darwinian Jericho.

The fossil record has thus not produced ancestors nor transitional forms for the major fish classes. Such hypothetical ancestors and the required transitional forms must, on the basis of the known record, be merely the products of speculation. How then can it be argued that the explanation offered by the evolution model to explain such evidence is more scientific than that of the creation model? In fact, the evidence *required* by evolution theory cannot be found.

SUMMARY

In summary, it can be said that the fossil record reveals an explosive appearance of highly complex forms of life without evidence of evolutionary ancestors. This fact is a great mystery to evolutionists; but creationists ask, *what greater evidence for creation could the rocks give than this sudden outburst of highly complex forms of life?* Furthermore, the fossil record fails to produce transitional forms between the major invertebrate types, between invertebrates and vertebrates, and between the major fish classes. The rocks cry out, "Creation!"

REFERENCES

1. G. G. Simpson, in *Evolution after Darwin*, Vol. 1. *The Evolution of Life*, ed. by Sol Tax, University of Chicago Press, Chicago, 1960, p. 143.
2. P. Cloud, *Geology*, Vol. 1, p. 123 (1973).
3. D. Axelrod, *Science,* Vol. 128, p.7 (1958).
4. P. Cloud, *Science*, Vol. 148, p. 27 (1965).
5. M. N. Bramlette, *Science*, Vol. 158, p. 673 (1967).
6. W. H. Bradley, *Science*, Vol. 160, p. 437 (1968).
7. A. E. J. Engel, B. Nagy, L. A. Nagy, C. G. Engel, G. O. W. Kremp, and C. M. Drew, *Science*, Vol. 161, p. 1005 (1968).

8. T. N. George, *Science Progress*, Vol. 48, p. 1 (1960).
9. G. G. Simpson, *The Meaning of Evolution*, Yale University Press, New Haven, 1949, p. 18.
10. E. G. Conklin, as quoted by G. E. Allen, *The Quarterly Review of Biology*, Vol. 44, p. 173 (1969).
11. A. S. Romer, *Vertebrate Paleontology*, 3rd edition, University of Chicago Press, Chicago, 1966, p. 12.
12. F. D. Ommaney, *The Fishes*, Life Nature Library, Time-Life, Inc., New York, 1964, p. 60.
13. E. White, *Proceedings Linnaen Society of London*, Vol. 177, p. 8 (1966).

Chapter V

THE FOSSIL RECORD — FROM FISH TO MAMMALS

NO RECORD OF FIN-TO-FEET TRANSITION

According to the assumed evolutionary sequence of life, fish gave rise to amphibia. This change would have required many millions of years and would have involved a vast multitude of transitional forms.

The fossil record has been diligently searched for a transitional series linking fish to amphibian, but as yet no such series has been found. The closest link that has been proposed is that allegedly existing between rhipidistian crossopterygian fish and amphibians of the genus, *Ichthyostega* (See Fig. 2). There is a tremendous gap, however, between the crossopterygians and the ichthyostegids, a gap that would have spanned many millions of years and

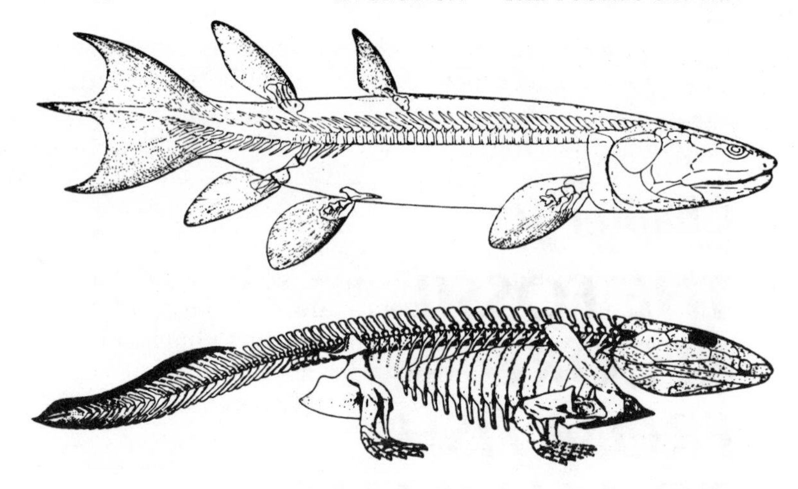

FIGURE 2. Reconstructions of an ichthyostegid amphibian and his supposed crossopterygian ancestor.

during which innumerable transitional forms should reveal a slow gradual change of the pectoral and pelvic fins of the crossopterygian fish into the feet and legs of the amphibian, along with loss of other fins, and the accomplishment of other transformations required for adaptation to a terrestrial habitat.

What are the facts? *Not a single transitional form has ever been found showing an intermediate stage between the fin of the crossopterygian and the foot of the ichthyostegid.* The limb and the limb girdle of *Ichthyostega* were already of the basic amphibian type, showing no vestige of a fin ancestry.

There is a basic difference in anatomy between all fishes and all amphibians not bridged by transitional forms. In all fishes, living or fossil, the pelvic bones are small and loosely embedded in muscle. There is no connection between the pelvic bones and the vertebral column. None is needed. The pelvic bones do not and could not support the weight of the body. There are no walking fishes, including the "walking catfish" of Florida. These latter fish do not walk, but slither along on their belly, using the same type of motion they use in the water.

In tetrapod amphibians, living or fossil, on the other hand, the pelvic bones are very large and firmly attached to the vertebral column. This is the type of anatomy an animal must have to walk. It is the type of anatomy found in all living or fossil tetrapod amphibians but which is absent in all living or fossil fishes. There are no transitional forms.

For a long time it was assumed that the fish that evolutionists believe gave rise to the amphibians became extinct about 70 million years ago. In rocks which evolutionists assume are 70 million years or younger, no fossils of the fish have ever been found. In about 1939, however, this type of fish was found to be alive and well off the coast of Africa. It is a crossopterygian fish of the genus *Latimeria.* It was taken from a depth of about 5,000 feet. Here he is still very much the same fish that is supposed to have given rise to the amphibians multiplied millions of years ago. It would certainly be astounding to believe that he has remained so genetically and morphologically stable for all those millions of years while his cousin was evolving all the way to man! Furthermore, how could *any* creature be on this earth for 70 million years without leaving a trace in the fossil record? Perhaps there is something wrong with evolutionary assumptions!

Why have rhipidistian crossopterygian fish been chosen as the ancestors of the amphibians? First of all, *there is nothing better available.* Lacking a candidate intermediate between fish and amphibian, evolutionists searched the various fish groups. The crossopterygians were then adopted as the fish group most likely to have been ancestral to amphibians. This was mainly because of certain skull patterns similar to those possessed by *Ichthyostega*, the possession of the complex "arch" type vertebrae similar to those found in *Ichthyostega* and other labyrinthodonts, and the presence of bones in the fins from which evolutionists assume it was possible for the amphibian limbs to have evolved.

According to Romer, the selective pressure which allowed the origin of tetrapod amphibians from their fish ancestors is the assumption that periodic droughts were characteristic of the Devonian Period, during which time the amphibians are supposed to have evolved. This forced the crossopterygian fish out of drying lakes and streams (they were believed to have possessed lungs) to seek other sources of water. Those forms which had inherited random mutational changes permitting more efficient locomotion on land would have survived in larger numbers than less well-equipped forms. A multitude of such episodes eventually gave rise, over many millions of years, to a true amphibian.[1]

This story, attractive at first sight, loses its plausibility when all the facts are considered. Since the amphibians are found in the later Devonian they would have had to evolve well back in the Devonian when their supposed crossopterygian ancestors were flourishing. If Romer's story is right, the Devonian should instead show mass extinctions of the crossopterygians and other fresh water forms. The opposite is true.

In the early Mississippian, three other amphibian orders are found. Since these highly diversified amphibians first appeared in the early Mississippian with ordinal characteristics already complete, they would have had to evolve well back in the Devonian, just as did, supposedly, the ichthyostegids. The same selection pressures, such as the periodic droughts postulated by Romer, must have been exerted on the antecedents of these three orders as on ichthyostegids, and they must have evolved from the crossopterygians or from the ichthyostegids. *But the members of none of these orders had the "arch" type vertebrae possessed by the crossopterygians and ichthyostegids*, but all possessed the "more primitive" lepospondylous, or "husk" type, vertebrae. How can the "arch" type vertebrae, then, be used to link the crossopterygians to amphibians?

Furthermore, of these three orders, the Aistopoda possessed long snake-like bodies, with up to 200 vertebrae. Most possessed no limbs whatsoever, not even a trace of a pectoral or pelvic girdle! In some forms of the order Nectridea the bodies were also elongate as in aistopods, with limbs likewise completely missing. If *Ichthyostega* or an ichthyostegid-like form was the ancestor of all amphibians, then while he was in the process of deriving tetrapod limbs from his crossopterygian ancestor, his amphibian offspring among the Aistopoda and Nectridea were just as busy finding a way to get rid of them!

What selective pressures gave rise to the tetrapod limb in the ichthyostegids while simultaneously causing its reduction and loss among aistopods and nectrideans? Why do these highly diverse forms appear in the fossil record with diversification already complete at their first appearance with *no evidence of transitional forms?*

The three living orders of amphibians include the salamanders and newts (Urodela or Caudata); the apodans (Apoda or Caecilia), earthworm-like with no limbs; and the frogs and toads (Anura or Salientia), which are among the most highly specialized of land vertebrates, possessing long hind limbs but bearing no tails. All of these modern amphibians possess the "more primitive" lepospondylous type vertebrae rather than the "arch" type vertebrae which supposedly link the amphibians to crossopterygian ancestors. Furthermore, *there are no transitional forms* linking these three modern orders, constituting the subclass Lissamphibia, and the amphibians found in the Paleozoic. Referring to the Lissamphibia, Romer states, "Between them and the Paleozoic group is a broad evolutionary gap not bridged by fossil materials."[2] Neither has the fossil record produced any evidence that these living amphibian orders have arisen from a common ancestor.

The extremely broad gap between fish and amphibia, as observed between the rhipidistian crossopterygians and the ichthyostegids, the sudden appearance, in fact, of all Paleozoic amphibian orders with diverse ordinal characteristics complete in the first representatives, and the absence of any transitional forms between these Paleozoic orders and the three living orders, makes it absolutely incredible to believe that these forms evolved. These facts, however, are completely in accord with the predictions of the creation model.

THE ALLEGED AMPHIBIAN-REPTILIAN-MAMMALIAN TRANSITION

It is at the amphibian-reptilian and the reptilian-mammalian boundaries that strongest claims have been advanced for transitional types bridging classes. But these are just those classes that are most closely similar in skeletal features, parts that are preserved in the fossil record.

The conversion of an invertebrate into a vertebrate, a fish into a tetrapod with feet and legs, or a nonflying animal into a flying animal are a few examples of changes that would require a revolution in structure. Such transformations should provide readily recognizable transitional series in the fossil record if they occurred through evolutionary processes. On the other hand, if the creation model is the true model, it is at just such boundaries that the absence of transitional forms would be most evident.

The opposite is true at the amphibian-reptilian and reptilian-mammalian boundaries, particularly the former. While it is feasible to distinguish between living reptiles and amphibians on the basis of skeletal features, they are much more readily distinguishable by means of their soft parts and, in fact, the major definitive characteristic which separates reptiles from amphibians is the possession by the reptile, in contrast to the amphibian, of the amniote egg.

Many of the diagnostic features of mammals, of course, reside in their soft anatomy or physiology. These include their mode of reproduction, warm-bloodedness, mode of breathing due to possession of a diaphragm, suckling of the young, and possession of hair.

The two most easily distinguishable osteological differences between reptiles and mammals, however, have never been bridged by transitional series. All mammals, living or fossil, have a single bone, the dentary, on each side of the lower jaw, and all mammals, living or fossil, have three auditory ossicles or ear bones, the malleus, incus, and stapes. In some fossil reptiles the number and size of the bones of the lower jaw are reduced compared to living reptiles. Every reptile, living or fossil, however, has at least four bones in the lower jaw and only one auditory ossicle, the stapes.

There are no transitional forms showing, for instance, three or two jaw bones, or two ear bones. No one has explained yet, for that matter, how the transitional form would have managed to chew while his jaw was being unhinged and rearticulated, or how he would hear while dragging two of his jaw bones up into his ear.

Furthermore, in order for the facts of the fossil record to fit the predictions of the evolution model, a true time-sequence must be established that accords with these predictions. This has not been possible with the amphibian-reptilian-mammalian sequence on the basis of fossil material so far discovered.

The known forms of *Seymouria* and *Didactes*, which are said to stand on the dividing line between amphibians and reptiles, are from the early Permian. This is at least 20 million years too late, according to the evolutionary time scale, to be the ancestors of the reptiles. The so-called "stem reptiles," from the order Cotylosauria, are found, not in the Permian or later, but in the preceding period, the Pennsylvanian.

In fact, the "mammal-like" reptiles of the suborder Synapsida, which are supposed to have given rise to mammals, are found in the Pennsylvanian, even possibly the early Pennsylvanian. Thus, *Seymouria* and *Didactes*, the "ancestors" of reptiles, would not only postdate the reptiles by tens of millions of years, but would postdate even the "ancestors" of mammals by an equal length of time.

According to evolutionists, mammals assumed supremacy over the reptiles at a relatively late time in reptilian history. If this is true, a reasonable assumption would be that the reptilian branch from which they arose developed late in the history of reptiles. Just the opposite is true, however, if the synapsids gave rise to mammals. The subclass Synapsida is dated among the earliest of known reptile groups, not the latest, and are supposed to have passed their peak even long before the appearance of dinosaurs.

According to Romer, the synapsid reptiles dwindled in numbers during the Triassic, becoming essentially extinct by the close of that period, and many millions of years elapsed before their mammalian "descendants" rose to a position of dominance.[3] If natural selection is one of the governing processes of evolution, and natural selection is defined as that process which enables the more highly adapted organism to produce the most offspring, then the above history, if true, seems to indicate that the reptile-to-mammal transition succeeded in spite of, rather than because of, natural selection.

The evidence for an amphibian-reptilian-mammalian transformation is thus much more the product of wishful thinking than that which can be adduced from the fossil record.

THE ORIGIN OF FLIGHT — AN EVOLUTIONARY ENIGMA

The origin of flight should provide an excellent test case for choosing between the evolution and creation models. Almost every structure in a nonflying animal would require modification for flight, and resultant transitional forms should be easily detectable in the fossil record. Furthermore, flight is supposed to have evolved four times separately and independently — in insects, birds, mammals (the bats), and in reptiles (the pterosaurs, now extinct). In each case the origin of flight is supposed to have required many millions of years, and almost innumerable transitional forms would have been involved in each case. *Yet not in a single case can anything even approaching a transitional series be produced.*

E. C. Olson, an evolutonist and geologist, in his book, *The Evolution of Life,*[4] states that, "As far as flight is concerned, there are some very big gaps in the record" (p. 180). Concerning insects, Olson says, "There is almost nothing to give any information about the history of the origin of flight in insects" (p. 180). Concerning flying reptiles, Olson reports that, "True flight is first recorded among the reptiles by the pterosaurs in the Jurassic Period. Although the earliest of these were rather less specialized for flight than the later ones, *there is absolutely no sign of intermediate stages*" (p. 181). (Emphasis added.) With reference to birds, Olson refers to *Archaeopteryx* as "reptile-like" but says that in possession of feathers "*it shows itself to be a bird*" (p. 182). (Emphasis added.) Finally, with reference to mammals, Olson states that, "The first evidence of flight in mammals is in *fully developed*

FIGURE 3. Restoration of two of the earliest birds, of the genus *Archaeopteryx*. Lull, R. S.: Organic Evolution, The Macmillan Co.

bats of the Eocene epoch" (p. 182). (Emphasis added.)

Thus, in not a single instance concerning origin of flight can a transitional series be documented, and in only one case has a single intermediate form been alleged. In the latter case, the so-called intermediate is no real intermediate at all because, as paleontologists acknowledge, *Archaeopteryx* was a true bird — it had wings, it was completely feathered, it *flew* (see Fig. 3). It was not a half-way bird, it *was* a bird.

Gregory has stated, "But in *Archaeopteryx*, it is noted, the feathers differ in no way from the most perfectly developed feathers known to us."[5] In reference to *Archaeopteryx, Ichthyornis,* and *Hesperornis*, Beddard stated: "So emphatically were all these creatures birds that the actual origin of Aves is barely hinted at in the structure of these remarkable remains."[6] During the 80 years since publication of Beddard's book, no better candidate as an intermediate between reptiles and birds than *Archaeopteryx* has appeared. Not a single intermediate with part-way wings or part-way feathers has been discovered. Perhaps this is why, with the passage of time, *Archaeopteryx*, in the eyes of some evolutionists, has become more and more "reptile-like"!

The alleged reptile-like features of *Archaeopteryx* consist of the claw-like appendages on the leading edges of its wings, the possession of teeth, and vertebrae that extend out along the tail. It is believed to have been a poor flyer, with a small keel or sternum. While such features might be expected if birds had evolved from reptiles, in no sense of the word do they constitute proof that

Archaeopteryx was an intermediate between reptile and bird. For example, there is a bird living today in South America, the hoatzin (*Opisthocomus hoatzin*), which in the juvenile stages possesses two claws.[7] Furthermore, it is a poor flyer, with an astonishingly small keel. This bird is unquestionably 100% bird, yet it possesses two of the characteristics which are used to impute a reptilian ancestry to *Archaeopteryx*!

The hoatzin is not the only living bird that possesses claws. The young of the touraco (*Touraco corythaix*; family Musophagidae) of Africa possesses claws and also is a poor flyer.[8] If either the hoatzin or touraco were found as fossils in appropriate strata, they would be hailed by evolutionists as transitional forms between reptiles and birds. But they are birds, living today! Furthermore, the ostrich has three claws on its wings, which, if one chose to do so, could be claimed to be more reptile-like than those in *Archaeopteryx*.

While modern birds do not possess teeth, some ancient birds possessed teeth, while some others did not. Does the possession of teeth denote a reptilian ancestry for birds, or does it simply prove that some ancient birds had teeth while others did not? Some reptiles have teeth while some do not. Some amphibians have teeth, but some do not. In fact, this is true throughout the entire range of the vertebrate subphylum — fishes, Amphibia, Reptilia, Aves, and Mammalia, inclusive.

Following the analogy that toothed birds are primitive while toothless birds are more advanced, the Monotremata (the duck-billed platypus and spiny anteater), mammals which do not possess teeth, should be considered more "advanced" than humans! Yet in every other respect these egg-laying mammals are considered to be the "most primitive" of all mammals. They did not appear, by the way, until the Pleistocene, which, on the evolutionary time-scale, makes them about 150 million years too late to be the ancestor of mammals! Just what evolutionary significance, then, can be assigned to the possession or absence of teeth?

Concerning the status of *Archaeopteryx*, duNouy, an evolutionist, has stated:

> Unfortunately, the greater part of the fundamental types in the animal realm are disconnected from a paleontological point of view. In spite of the fact that it is undeniably related to the two classes of reptiles and birds (a relation which the anatomy and physiology of *actually living specimens* demonstrates), we are not even authorized to consider the exceptional case of the *Archaeopteryx* as a true link. By link, we mean a necessary stage of transition between classes such as reptiles and birds, or between smaller groups. An animal displaying characters belonging to two different groups cannot be treated as a true link as long as the intermediary stages have not been found, and as long as the mechanisms of transition remain unknown.[9] (Emphasis added.)

Swinton, an evolutionist and an expert on birds, states: "The origin of birds is largely a matter of deduction. There is no fossil evidence of the stages through which the remarkable change from reptile to bird was achieved."[10] Thus, most emphatically, the fossil record does not document the assumed transition from reptile to bird, but birds appear abruptly in the fossil record, just as predicted on the basis of creation.

In fact, a very recent discovery completely demolishes *Archaeopteryx* as a transitional form. A note in *Science News*, Vol. 112, Sept. 24, 1977, p. 128, made the startling announcement (startling to evolutionists, that is) that a fossil of an undoubted true bird has been found in rocks of the same geological period as *Archaeopteryx*! This means, according to Professor John Ostrom of Yale University, that formations, much older than those containing *Archaeopteryx* (the Jurassic) will have to be searched for the ancestor of birds. Obviously, *Archaeopteryx* cannot be the ancestor of birds if true birds existed at the same time. *Archaeop-*

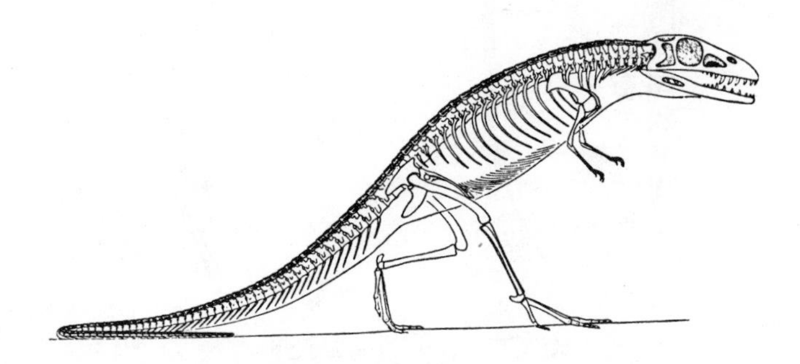

FIGURE 4. *Saltoposuchus*, a thecodont, believed by Romer and others to be the ancestor of dinosaurs, birds, and reptiles. From Romer's *Vertebrate Paleontology*, by permission of The University of Chicago Press.

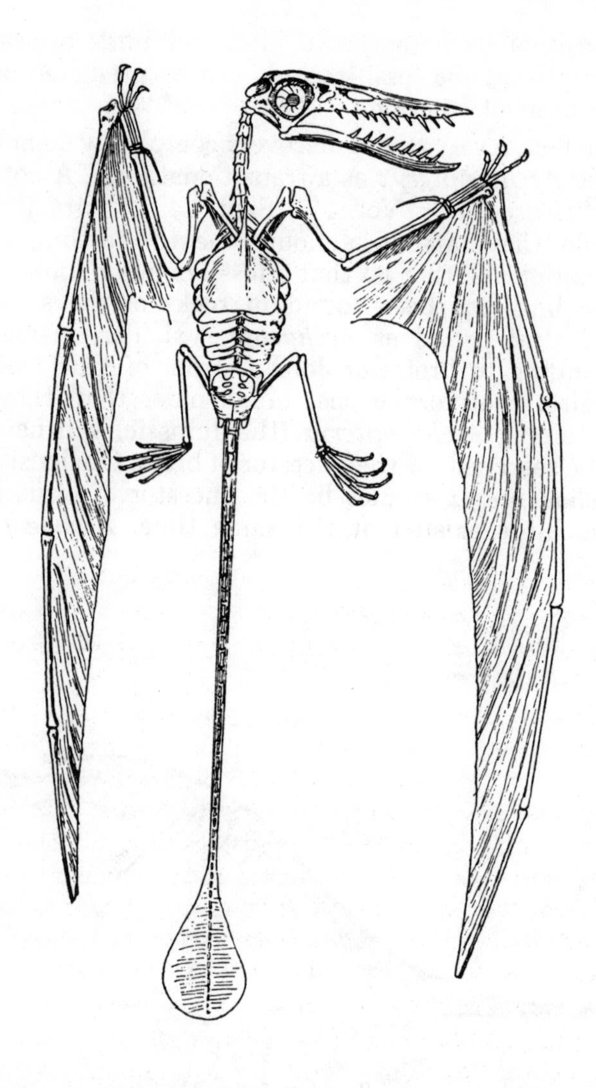

FIGURE 5. *Rhamphorhynchus*, a long-tailed pterosaur. From Williston's *The Osteology of the Reptiles*, by permission of the Harvard University Press.

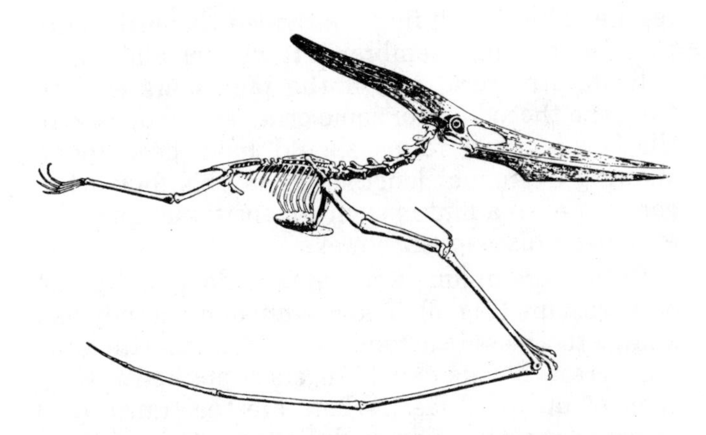

FIGURE 6. *Pteranodon*, a giant flying reptile with a wingspan of over 50 feet. From Romer's *Vertebrate Paleontology*, by permission of The University of Chicago Press.

teryx, the one example that evolutionists have repeatedly insisted is the best example of a transitional form, is thus absolutely destroyed as a transitional form.

The differences between nonflying and flying reptiles were especially dramatic. In Figure 4 is shown a reconstruction of *Saltoposuchus*, a representative of the thecodonts that Romer believes gave rise to flying reptiles (pterosaurs), dinosaurs, and birds. The vast gulf between this creature and *Archaeopteryx* is obvious. Equally obvious is the tremendous gap between *Saltoposuchus* and representatives of the two suborders of the pterosaurs, shown in Figures 5 and 6.

Almost every structure in *Rhamphorhynchus*, a long-tailed pterosaur (Fig. 5), was unique to this creature. Especially obvious (as in all pterosaurs) was the enormous length of the fourth finger, in contrast to the other three fingers possessed by this reptile. This fourth finger provided the entire support for the wing membrane. It was certainly not a delicate structure, and if the pterosaurs evolved from the thecodonts or some other earthbound reptile, transitional forms should have been found showing a gradual lengthening of this fourth finger. Not even a hint of such a transitional form has ever been discovered, however.

Even more unique was the pterodactyloid group of pterosaurs (Fig. 6). The *Pteranodon* not only had a large toothless beak and long rearward-extending bony crest, but its fourth fingers supported a wingspan of up to 52 feet. Where are the transitional forms documenting an evolutionary origin of these and other structures unique to the pterosaurs? How could these strange creatures have evolved through innumerable intermediate forms over millions of years of time without leaving a single such intermediate in the fossil record? The answer is, they did not evolve — they were created!

The bat is presumed to have evolved from an insectivore (insectivores include such creatures as the moles, shrews, and hedgehogs), a nonflying mammal. A fossil bat is shown in Figure 7. Truly, a revolution in structure would be required to convert an animal such as a mole, hedgehog, or shrew into a bat. In the bat, four of the five fingers are extremely long compared to the normal hand and support the wing membrane. If bats evolved from an insectivore or some other creature, transitional forms should be found documenting the origin of these and other structures unique to the bat.

As already noted, no such intermediate forms have been found anywhere in the fossil record. In Figure 8 is shown what is supposed to be the oldest known bat. This fossil bat was found in rocks allegedly 50 million years old. The article accompanying this picture stated that no fossil related to a bat any older than this has ever been found. The picture (Fig. 8) includes a photograph of the bones of the bat as well as a reconstruction of what this bat would have looked like.

Well, here he is — the world's oldest known bat. And what is he? One hundred percent bat! The complete absence of any supposed transitional forms between the bat and his alleged ancestor leaves unanswered, on the basis of the evolutionary hypothesis, such questions as when, from what, where, and how did bats originate?

Now let us ask the question: "Concerning the origin of flight, which model, the creation model or the

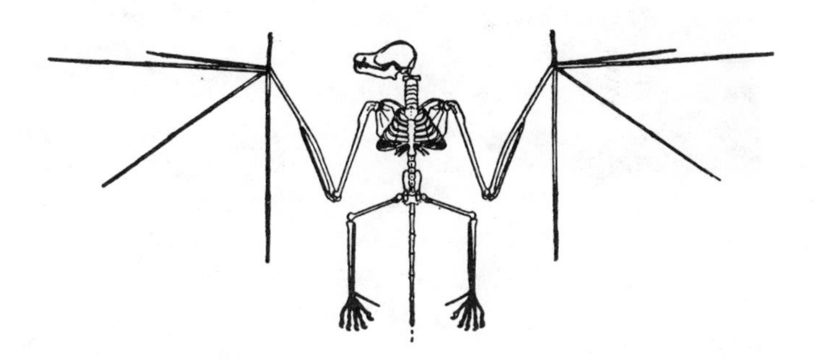

FIGURE 7. Skeleton of a fossil bat, *Palaeochiropteryx*. From Romer's *Vertebrate Paleontology*, by permission of The University of Chicago Press.

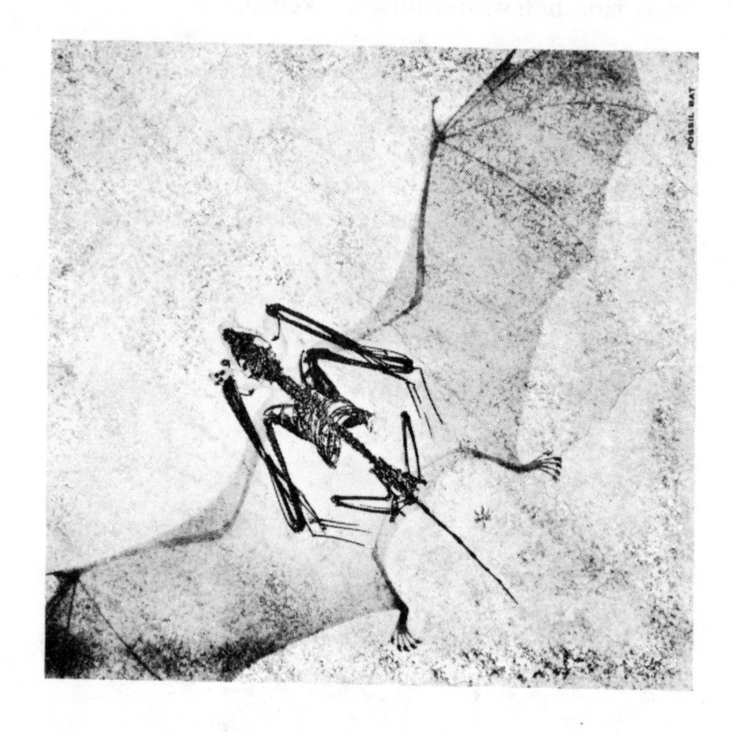

FIGURE 8. Photo of the oldest fossil bat superimposed on its reconstruction, by G. L. Jepsen. From the cover page of *Science*, December 9, 1966. Copyright 1966 by the American Association for the Advancement of Science.

evolution model, has greater support from the fossil record?" To us the answer seems obvious. Not a single fact contradicts the predictions of the creation model, but the actual evidence fails miserably to support the predictions of the evolution model. Here, where transitional forms should be the most obvious and easy to find — if evolution *really* accounted for the origin of these highly unique creatures — *none* are found. Could the fossil record really be that cruel and capricious to evolutionary paleontologists? The historical record inscribed in the rocks literally cries, "Creation"!

RODENTS — THE MOST PROLIFIC MAMMAL — PRODUCE NO EVIDENCE FOR EVOLUTION

The order Rodentia should provide evolutionists with another group of animals ideal for evolutionary studies. In number of species and genera, the rodents exceed all other mammalian orders combined. They flourish under almost all conditions. Surely, if any group of animals could supply transitional forms, this group could.

As to their origin, Romer has said:

> The origin of the rodents is obscure. When they first appear, in the late Paleocene, in the genus *Paramys*, we are already dealing with a typical, if rather primitive, true rodent, with the definitive ordinal characters well developed. Presumably, of course, they had arisen from some basal, insectivorous, placental stock; but *no transitional forms are known.*[11] (Emphasis added.)

Furthermore, transitional forms between the basic rodent types are not found in the fossil record. For example, Romer says:

> . . . the beavers are presumably derived from some primitive sciuromorph stock, but there are no annectant types between such forms and the oldest Oligocene castoroids to prove direct relationship.

Speaking of the Hystricidae, the Old World porcupines, Romer says:

> There are a few fossil forms, back to the Miocene and possibly late Oligocene, but these give no indication of relationship of hystricids to other rodent types.

Commenting on the "rock rat," *Petromus*, Romer says, "Almost nothing is known of the ancestry of *Petromus*." Of the superfamily Theridomyoidea, Romer says, "At present we know nothing of their ancestry or possible descendants." Of the lagomorphs (hares and rabbits), once placed in a suborder of the rodents, but now placed in a separate order, Lagomorpha, Romer must admit that, "The lagomorphs show no close approach to other placental groups, and the ordinal characters are well developed in even the oldest known forms."

Thus we see that the order Rodentia, which should supply an excellent case for evolution, if evolution really did occur, offers powerful evidence against the evolutionary hypothesis.

REFERENCES

1. A. S. Romer, *Vertebrate Paleontology*, 3rd Edition, University of Chicago Press, Chicago, 1966, p. 36.
2. Ibid., p. 98.
3. Ibid., p. 173.
4. E. C. Olson, *The Evolution of Life*, The New American Library, New York, 1965.
5. W. K. Gregory, *New York Academy of Science Annals*, Vol. 27, p. 31 (1916).
6. F. E. Beddard, *The Structure and Classification of Birds*, Longmans, Green and Co., London, 1898, p. 160.
7. J. L. Grimmer, *National Geographic*, p. 391 (September 1962).
8. C. G. Sibley and J. E. Ahquist, *Auk*, Vol. 90, p. 1 (1973).
9. L. du Nouy, *Human Destiny*, The New American Library, New York, 1947, p. 58.
10. W. E. Swinton, in *Biology and Comparative Physiology of Birds*, ed. by A. J. Marshall, Academic Press, New York, Vol. 1, 1960, p. 1.
11. A. S. Romer, Ref. 1, p. 303.

Chapter VI

THE ORIGIN OF MAN

The situation is no different with the order Primates, within which man has been placed (see Fig. 9). The prosimians are supposed to be the earliest representatives of the primates, and evolutionists thus believe that all other primates have evolved from one of these prosimians. The living prosimians include the lemurs, lorises, and tarsiers.

Although the primates are supposed to have evolved from an insectivorous ancestor, there are no series of transitional forms connecting primates to insectivores. Elwyn Simons, one of the world's leading experts in the field of primates, must admit that, "In spite of recent finds, the time and place of origin of order Primates remains *shrouded in mystery*."[1] (Emphasis added.) Romer states that the early lemurs appear "apparently as immigrants from some unknown area."[2] He is forced to say this since paleontologists simply cannot tell from the fossils how lemurs arose. Kelso has stated, ". . . the transition from insectivore to primate is not documented by fossils. The basis of knowledge about the transition is by inference from living forms."[3] There are no transitional forms found anywhere in the fossil record.

ORDER : PRIMATES

SUBORDER : PROSIMII

LEMURS, LORISES, TARSIERS

SUBORDER : ANTHROPOIDEA

Platyrrhines

NEW WORLD MONKEYS

Catarrhines

OLD WORLD MONKEYS

APES

MAN

FIGURE 9. Classification of the order Primates.

We can see, then, at the very outset the true origin of the entire primate order cannot be determined from the fossil record. If primates evolved, there should be a series of transitional forms leading back to their insectivore ancestors, but no such transitional forms are found. This is exactly what creationists would expect the record to show, of course.

The prosimians are supposed to have given rise to the platyrrhines (South American or New World monkeys) and the catarrhines, which include the Old World monkeys, apes, and man. There are no transitional forms between the South American monkeys and their presumed ancestors, the prosimians. Thus Romer states, "Little is known, unfortunately, of the fossil history of the South American monkeys."[4] Kelso states:

> The details of the evolutionary background of the New World monkeys, the Platyrrhinae, would doubtless be informative and interesting, but unfortunately we know very little about them.[5]

Very little indeed! Nothing in fact is known. When monkeys first appear in South America they are just that — monkeys. No transitional forms are available. Neither has anyone succeeded in finding the ancestors of the Old World monkeys. Thus Kelso states:

> Clearly, the fossil documentation of the emergence of the Old World monkeys could provide key insights into the general evolutionary picture of the primates, but, in fact, this record simply does not exist.[6]

There are simply no transitional forms between the prosimians and the catarrhines. Thus Simons frankly states, "Although the word has been used, there is actually no such thing as a 'protocatarrhine' known from the fossil record."[7] In a later publication he said:

> . . . not a single fossil primate of the Eocene epoch from either continent* appears to be an acceptable ancestor for the great infraorder of the catarrhines, embracing all of the living higher Old World primates, man included.[8]

Already, then, the fossil record has failed twice to produce man's supposed progenitors: antecedents of the entire primate order are lacking and transitional forms between the prosimians, allegedly the more "primitive" of the primates, and the catarrhines, or more "advanced" primates, have never been found.

Romer refers to the chimpanzee, *Pan,* and the gorilla, *Gorilla*, as the "highest living members of the anthropoid group." What can he say about their origin? Romer states, "Our knowledge of the fossil history of these higher apes and of presumed human ancestors on this level is tantalizingly poor."[9] Some have imagined that the ancestors of the chimpanzee, gorilla, and orangutan may be found among species of *Dryopithecus,* fossil apes found in Africa, Europe, and Asia.[10] Just why this is so, seems to be rather obscure, to say the least.

What can anthropologists say about man's ultimate origin from his imagined ape-like ancestor? Pilbeam says:

> It has come to be rather generally assumed, albeit in a rather vague fashion,

---

*These "early" prosimians are found only in North America and Europe.

the pre-Pleistocene hominid ancestry was rooted somewhere in the Dryopithecinae.[10]

When a scientist is forced to "assume" something "in a rather vague fashion," it is obvious that he is resorting to wholly unscientific methods to establish what he cannot do by the usual accepted scientific method. What strange qualities could paleoanthropologists detect in an animal that allows them to decide on one hand that it was the progenitor of the chimpanzee, gorilla, and the orangutan, and yet on the other hand was the progenitor of the human race?

Pilbeam apparently does not agree with the general assumption that the dryopithecines were ancestral to man. He has expressed his conviction that the dryopithecines were too specialized, already too committed to ape-dom to have produced the hominids.[11] Evolutionists simply have not been able to find man's supposed ultimate ancestor. The imagined common ancestor of ape and man awaits to be discovered.

RAMAPITHECUS — AN EARLY HOMINID?

A creature that Pilbeam, Simons, and other evolutionists feel definitely was a hominid, or a creature with some man-like traits, was *Ramapithecus*. They consider that *Ramapithecus* was in the lineage leading from ape to man. *Ramapithecus* consists merely of a handful of teeth and jaw fragments.[1,8,10-12] On the basis of this extremely fragmentary evidence, some evolutionists have built an evolutionary ancestor for man.

These evolutionists believe that *Ramapithecus*

was in man's lineage because the incisors and canine teeth (the front teeth) of this animal were relatively small in relation to the cheek-teeth (as is the case in man); they believe the shape of the jaw was parabolic, as in humans, rather than U-shaped, as in most apes; and because of some other subtle anatomical distinctions found for the jaw fragments. The face is also believed to have been foreshortened, although no bones of the face or skull have been recovered.

Thus, all of the evidence linking *Ramapithecus* to man is based solely upon extremely fragmentary dental and mandibular (jaw) evidence. Recent evidence apparently invalidates even this. Before we go further we must explain that the term "hominoid" refers to both hominids and pongids, or apes. Thus, both apes and man are called hominoids, while only man and "man-like apes" are called hominids.

Recently, Dr. Robert Eckhardt, a paleoanthropologist at Penn State, published an article[13] headlined by the statement:

> Amid the bewildering array of early fossil hominoids, is there one whose morphology marks it as man's hominid ancestor? If the factor of genetic variability is considered, the answer appears to be no.

In other words, according to Eckhardt, nowhere among the fossil apes or ape-like creatures can be found what could be judged to be a proper ancestor for man. As has been noted above, Simons, Pilbeam, and others consider *Ramapithecus* to have been a hominid, and this judgment has been made solely on the basis of a few teeth and a few fragments of the jaw. Eckhardt made 24 different measurements on a collection of fossil teeth from two species of *Dryopithecus* (fossil apes) and one species of *Ramapithecus* (a supposed fossil hominid) and compared the range of variation found for these fossil species to similar measurements made on a population of chimpanzees at a research center and on a sample of wild chimpanzees in Liberia.

The range of variation in the chimpanzee populations was actually greater than those in the fossil samples for 14 of the 24 measurements, the same for one, and less for 9 of the measurements. Even in the minority of cases where the range of variation of the fossil samples exceeded those in living chimpanzees, the differences were very small. Thus, in the tooth measurements made, there was greater variation among living chimpanzees, or a single group of apes, than there was between *Dryopithecus*, a fossil ape, and *Ramapithecus*, which is supposed to have been a hominid. And remember, *Ramapithecus*, was judged to be a hominid solely on the basis of its dental characteristics!

Eckhardt extended his calculations to 5 other species of *Dryopithecus* and to *Kenyapithecus*, which according to Simons and Pilbeam[10,14] is equivalent to *Ramapithecus*. After stating that on the basis of tooth-size calculations, there is little basis for classifying the dryopithecines in more than a single species, Eckhardt goes on to say:

> Neither is there compelling evidence for the existence of any distinct hominid species during this interval, unless the designation "hominid" means simply any individual ape that happens to have

small teeth and a corresponding small face.

Eckhardt's conclusion is that *Ramapithecus* seems to have been an ape — morphologically, ecologically, and behaviorly.

Even more devastating evidence against the assumption of a hominid status for *Ramapithecus* has been recent revelations concerning a living high-altitude baboon found in Ethiopia. This baboon, *Theropithecus galada*, has incisors and canines which are small relative to those of extant African apes, closely packed and heavily worn cheek teeth, powerful masticatory muscles, and short deep face, and other "man-like" features possessed by *Ramapithecus* and *Australopithecus*.[15,16] Since this animal is nothing but a baboon in every other feature, and living today, it is certain that it has no genetic relationship to man. Yet it has many of the dental and mandibular characteristics used to classify *Ramapithecus* as a hominid!

While it is true that the possession of such features by monkeys or apes is highly exceptional, to include these dental and mandibular characteristics among those considered to be diagnostic of hominids, since they *are* possessed at least in one case by a baboon, is obviously invalid. These facts would render highly uncertain, if not impossible, the classification of any fossil as a hominid solely on the basis of dental and associated characteristics.

These considerations and the information compiled by Eckhardt offer almost compelling evidence that *Ramapithecus* was no hominid at all, but was an ape or monkey with a diet and habitat similar to that of galada baboons. Thus there is no real evidence for a hominid of any kind in the huge gap between the supposed branching point of man and ape and the australopithecines, which we will consider shortly. Many evolutionists believe that man's ancestors branched off from the apes about 30 million years ago, and they date the australopithecine fossils from one to three million years or so. Based on their dating methods (which we consider invalid), this would mean that there was a period of about 25 million years during which hominids were supposedly evolving, yet not a single fossil hominid of that period has been discovered.

AUSTRALOPITHECUS — APE OR APE-MAN?

The next, and much more recent, candidate, chronologically speaking, as one of man's hominid ancestors, is *Australopithecus*.[17] The first find of this creature was by Dart in 1924, to which he gave the name *Australopithecus africanus*. He pointed out the many ape-like features of the skull, but he believed that some features of the skull and particularly of the teeth were man-like. The name *Australopithecus* means "southern ape," but after Dart examined the teeth further, he decided *A. africanus* was a hominid. This claim created considerable controversy, most workers at that time claiming that *A. africanus* was an ape with some interesting but irrelevant parallel features with man. Additional finds of *Australopithecus* were made in later years by Broom and by Dart.

The find by Louis Leakey and his wife of what they called *Zinjanthropus bosei*, or "East-Africa Man," at Olduvai Gorge in Tanzania has attracted great attention from the public. As it turned out, they really found nothing essentially different than had been discovered by Dart many years earlier. Their research, however, was sponsored by the National Geographic Society, and a combination of rather extravagant claims by Leakey for his find combined with publicity through the pages of the National Geographic Society magazine succeeded in conveying the idea that Leakey had made a unique and momentous discovery at Olduvai. Even Leakey admitted later, however, that his *Zinjanthropus bosei* is a variety of *Australopithecus,* discovered years previously in South Africa.

The australopithecines have been classified into two species. One is more gracile with somewhat smaller jaws and teeth and has been designated *Australopithecus africanus* (Fig. 10). The other has more massive teeth and jaws and possesses sagittal and supramastoid crests (bony ridges), found in gorillas and orangs, and has been named *Australopithecus robustus* (Fig. 11).

All of these animals possessed small brains, the cranial capacity averaging 500 c.c. or less, which is in the range of a gorilla, and about one-third of that for man. These animals thus unquestionably had the brains of apes, regardless of what else can be said about them. Both of them had ape-like skulls and jaws, these features being particularly obvious in the case of *A. robustus.*

The dentition, above all, it is said, is what makes these animals distinctive and which has served to cause paleoanthropologists to claim a hominid status for them (Fig. 12). The front teeth, incisors, and canines, are relatively small, and the dental arcade, or curve of the jaw, is more parabolic and less U-shaped than is typical of modern apes. It is also claimed that the morphology, or shape, of the

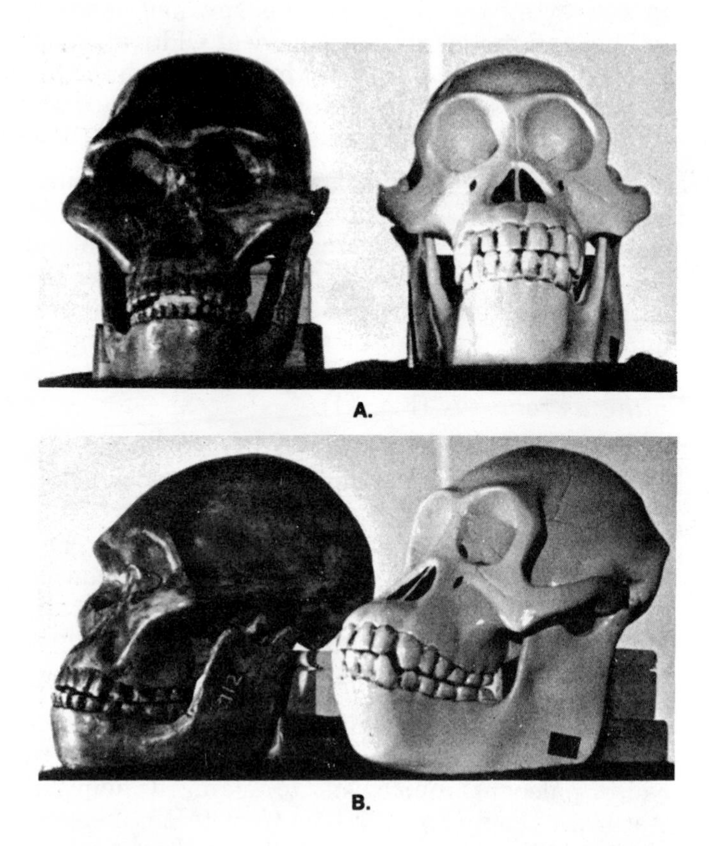

FIGURE 10. Frontal (A) and diagonal (B) views of Australopithecus africanus (left) and cast of an orangutan skull (right). From Rusch's *Human Fossils,* in *Rock Strata and the Bible Record,* P. A. Zimmerman, Ed., Concordia Publishing House.

FIGURE 11. Reconstruction of the *Australopithecus robustus (Zinjanthropus)* skull.

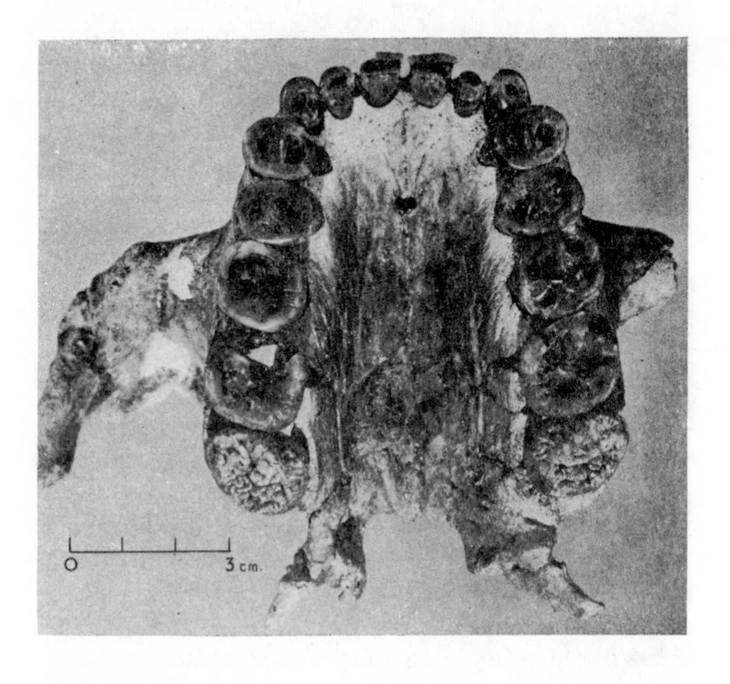

FIGURE 12. The palate and dentition of *Australopithecus robustus.*

teeth is in many features more man-like than ape-like. The cheek teeth (premolars and molars), however, are massive, even in the gracile, or africanus form. *A. africanus*, even though only about 60-70 lbs. or about the size of a smallish chimpanzee, had cheek teeth larger than chimps and orangs and as large as gorillas, some of the latter of which reach 400 lbs. in size. As a consequence, the jaws are very large, particularly in *A. robustus*.

Some fragments of the pelvis, limb, and foot bones of these animals have been recovered and, based on studies of these fragments, it has been the consensus among evolutionists that the australopithecines walked habitually upright. This was especially so after such authorities as Broom[18] and LeGros Clark[19] strongly supported this conclusion. In more recent years, however, this view has been challenged by Solly Lord Zuckerman,[20,21] famous British anatomist, and by Dr. Charles Oxnard, [22,23] professor of anatomy and anthropology at the University of Chicago.

For over 15 years a research team headed by Lord Zuckerman studied the anatomical features of man, monkeys, apes, and the australopithecine fossils. Practically all available important fossil fragments of *Australopithecus*, along with anatomical specimens from hundreds of monkeys, apes, and humans were compared. No one has done a more thorough and careful study on the status of *Australopithecus* than Lord Zuckerman.

Concerning the claim by LeGros Clark and others that *Australopithecus* should be classified as a genus of the Hominidae (family of man) rather than as a genus of the anthropoid apes, Lord Zuckerman said:

> But I myself remain totally unpersuaded. Almost always when I have tried to check the anatomical claims on which the status of *Australopithecus* is based, I have ended in failure.[24]

Lord Zuckerman's conclusion is that *Australopithecus* was an ape, in no way related to the origin of man.

Oxnard's research has led him to say:

> Although most studies emphasize the similarity of the australopithecines to modern man, and suggest, therefore, that these creatures were bipedal tool-makers at least one form of which (*Australopithecus africanus* — "*Homo habilis*," "*Homo africanus*") was almost directly ancestral to man, a series of multivariate statistical studies of various postcranial fragments suggests other conclusions.[25]

From his results Oxnard concluded that *Australopithecus* did not walk upright in human manner but probably had a mode of locomotion similar to that of the orang. He states:

> Multivariate studies of several anatomical regions, shoulder, pelvis, ankle, foot, elbow, and hand are now available for the australopithecines. These suggest that the common view, that these fossils are similar to modern man or that on those occasions when they depart from a similarity to man they resemble the African great apes, may be incorrect. Most of the fossil fragments are in fact uniquely different from both man and man's nearest living genetic relatives, the chimpanzee and gorilla.

Zinjanthropus drawn by Neave Parker for Dr. L. S. B. Leakey.
Copyright, The Illustrated London News & Sketch, Ltd., 9/1/60.

Zinjanthropus drawn by Maurice Wilson for Dr. Kenneth P. Oakley.
by kind permission of Dr. Kenneth P. Oakley.

FIGURE 13. Two contrasting views by evolutionists of *Zinjanthropus (A. robustus).*

> To the extent that resemblances exist with living forms, they tend to be with the orangutan.[26]

Oxnard's conclusions are, then, that *Australopithecus* is not related to anything living today, man or ape, but was uniquely different. If Oxnard and Lord Zuckerman are correct, certainly *Australopithecus* was neither ancestral to man nor intermediate between ape and man.

As noted earlier, the claim that the australopithecines were hominids and in man's lineage is based upon their dentition and the belief that they walked upright. As for their dentition, they did have relatively small front teeth, but their cheek teeth were broad and very large and their jaws were large and in some cases, massive in size. Furthermore, as already noted, a living baboon, *Theropithecus galada*, has a number of dental, mandibular, and facial characteristics which are shared with the australopithecines. This fact is particularly damaging to the dental evidence for a hominid status for the australopithecines.

Combining all of the above considerations with the undoubted fact that the australopithecines possessed ape-sized brains, strongly indicates, I believe, that they were nothing more than aberrant apes, ecologically similar, perhaps, to galada baboons. Even earlier than Zuckerman and Oxnard, some evolutionists had expressed their belief that the australopithecines were simply specialized apes. Thus Ashley Montagu, a well-known evolutionist, has stated that, ". . . the skull form of all australopithecines is extremely ape-like. . . . the australopithecines show too many specialized and ape-like characters to be either the direct ancestor of man or of the line that led to man."[27] It has been argued by Robinson[28] and by others[17] that *Homo habilis* is the same as *A. africanus*. If this is true, then the above arguments would also apply to this creature.

JAVA MAN

It is claimed by some that fragments of *Homo erectus* are found at Olduvai Gorge. The story of *Homo erectus* begins elsewhere, however. The creatures popularly called Java Man and Peking Man are now classified as a single species, namely, *Homo erectus*.

A Dutch physician by the name of Dubois became convinced that the "missing link" in man's evolutionary origin would be found in the East Indies. He joined the Dutch army and got an assignment to Java, where he began his search. In 1891, along the bank of the Solo River near the village of Trinil, he found a skullcap. The skullcap was very low-vaulted, (low, sloping forehead) with heavy brow ridges. The cranial capacity was estimated by Dubois to be about 900 c.c., about two thirds of that for modern man, but it is impossible to determine cranial capacity from a skullcap alone.

About a year later and 50 feet from where he had found the skullcap, Dubois found a human femur (thigh bone). Dubois assumed, without any real justification, that the skullcap and femur were from the same individual. He placed them together and called the association *Pithecanthropus erectus* ("erect ape-man"). At about the same time, Dubois found two molar teeth which he included in his original announcement. In 1898 he discovered a premolar tooth which he believed should also be included with the first finds. The collection came to be known as Java Man. Evolutionists have estimated the age of the fossils to be about 500,000 years.

Dubois exhibited these fossils at the International Congress of Zoology at Leyden in 1895. Authorities greeted Dubois' announcement with considerable skepticism and divided opinion. British zoologists tended to view the remains as human, the Germans as those of an ape, and the French as those of something between ape and man.

Dubois concealed the fact that he had also discovered at nearby Wadjak and at approximately the same level two human skulls (known as the Wadjak skulls) with a cranial capacity of about 1550-1650 c.c., somewhat above the present human average. To have revealed this fact at that time would have rendered it difficult, if not impossible, for his Java Man to have been accepted as a "missing link." It was not until 1922, when a similar discovery was about to be announced, that Dubois revealed the fact that he had possessed the Wadjak skulls for over 30 years. His failure to reveal this find to the scientific world at the same time he exhibited the *Pithecanthropus* bones can only be labeled as an act of dishonesty and calculated to obtain acceptance of *Pithecanthropus* as an ape-man.

Before his death and after he had convinced most evolutionists as to the man-like affinity of *Pithecanthropus*, Dubois himself changed his mind and declared that his Java Man was nothing more than a large gibbon![29]

Boule and Vallois (Boule before his death was the Director of the French Institute of Human Paleontology and one of the world's greatest authorities on fossil skulls), after a thorough discussion of the skullcap found by Dubois, said, "Taken as a whole, these structures are very similar to those of chimpanzees and gibbons."[30] They report that von Koenigswald ascribed the two molar teeth to an orang and the premolar to a true man.[31] In other words, the teeth were in no way associated with the original owner of the skullcap.

A 1906 expedition at the exact site of Dubois' excavation failed to produce a single scrap of similar material, although 10,000 cubic yards of soil were removed. During 1936-1939, G. H. R. von Koenigswald carried out an extensive search at Sangiran, about 40 miles from Trinil. His efforts were rewarded by the discovery of fragments of jawbones, including teeth, fragments of skulls, and a skullcap. No limb bones were found. Von Koenigswald labeled his finds *Pithecanthropus* II, III, and IV.

Boule and Vallois report that the skulls found at Sangiran present the same general characters as Dubois' *Pithecanthropus*,[30] which, as mentioned above, they had likened to those of chimpanzees and gibbons. In the case of the Sangiran finds, several teeth were intact in the mandible. Every characteristic of these teeth given by Boule and Vallois is simian rather than man-like.[31]

The assessment of the femur found by Dubois at Trinil (plus a few other fragments of femora found later by Dubois) was that it was indistinguishable from that of a human. They conclude:

> If we possessed only the skull and the teeth, we should say that we are dealing with beings, *if not identical with*, at least closely allied to the Anthropoids. If we had only the femora, we should declare

> we are dealing with Man.[32] (Emphasis added.)

It is clear, then, that the only justification for assessing *Pithecanthropus* as anything more than an ape was the association of the human femur with the other remains. There was no justification for associating the human femur with the ape-like skull, however, except that Dubois and other evolutionists needed a "missing link." If the two molars and the premolar collected at that time by Dubois turned out to be from an orang and a true man, respectively, why continue to associate the femur with the skull cap? Remember that Dubois, who knew more about the circumstances than anyone else, finally renounced *Pithecanthropus* and labeled the skullcap as that of a giant gibbon. We believe that the claim for a man-like status for *Pithecanthropus* should be laid to rest.

PEKING MAN

If one accepts uncritically the evidence usually presented in texts and treatises on Peking Man, the case for the existence of near-man, or a man with many very primitive features, would seem established. For example, the skull model and flesh reconstructions based on this model shown in Figure 14 reveal a remarkable resemblance to modern man and could hardly be called less than human. A close examination of the reports related to Peking Man, however, reveal a tangled web of contradictions, highly subjective treatment of the data, a peculiar and unnatural state of the fossil bones, and the loss of essentially all of the fossil material.

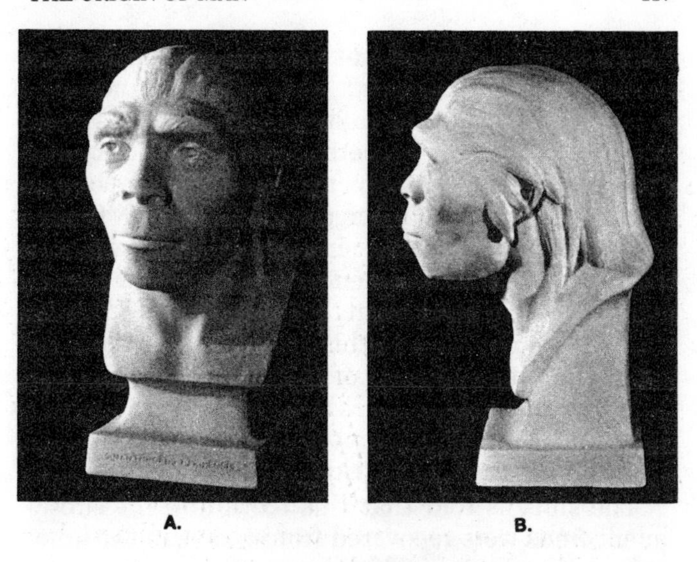

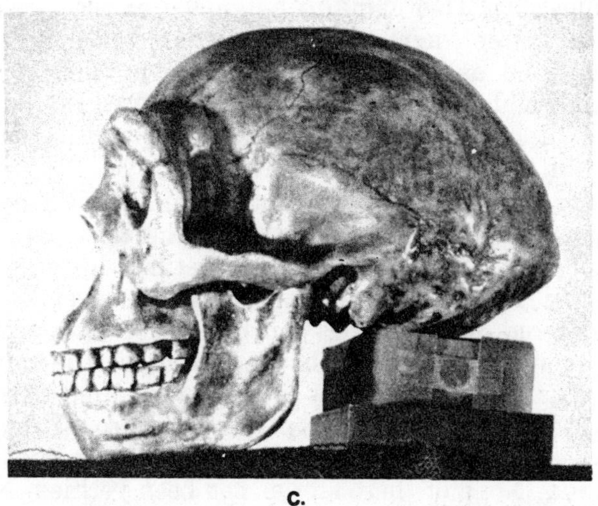

FIGURE 14. Flesh model (A, B) and skull model (C) of *Sinanthropus pekinensis* (so-called Peking Man). From Rusch's *Human Fossils*, in *Rock Strata and the Bible Record*, P. A. Zimmerman, Ed., Concordia Publishing House.

At Choukoutien, about 25 miles from Peking, China, in the 1920's and 1930's, were found fragments of about 30 skulls, 11 mandibles (lower jaws), and about 147 teeth. Except for a very few and highly fragmentary remains of limb bones, nothing else from these creatures was found. One of the initial finds was a single tooth, and without waiting for further evidence, Dr. Davidson Black, Professor of Anatomy at Union Medical College, Peking, declared that this tooth established evidence for the existence of an ancient hominid, or man-like creature, in China. He designated this creature *Sinanthropus pekinensis*, which soon came to be known as Peking Man.

The story is told that this tooth and the subsequent finds were recovered from a cave in the limestone cliff. This came to be known as the "lower cave" after fragments of ten other creatures, all identified as the remains of modern man, were found higher up on the cliff in what was allegedly the "upper cave." As we shall see, there is serious doubt that a cave existed at either level.

Of most critical importance to an evaluation of this material is the fact that all of this material except two teeth disappeared sometime during the period 1941-1945, and none of it has ever been recovered. Many stories concerning the disappearance of this material have circulated, the most popular being that it was either lost or seized by the Japanese during an attempt to move it from Peking to a U.S. marine detachment that was evacuating China. None of these stories has been verified. No living person apparently knows what happened to the material.

As a result we are totally dependent on models and descriptions of this material left by a few investigators, all of whom were evolutionists and totally committed to the idea that man had evolved from animal ancestors. Even if a scientist is completely honest and as objective as humanly possible, the model or description he fashions on the basis of scanty and incomplete material will reflect to a critical degree what he thinks the evidence ought to show. Furthermore, there is ample evidence that objectivity was seriously lacking in the treatment and evaluation of the material recovered at Choukoutien. If the type of evidence we have today relating to Peking Man was brought into a court of law, it would be ruled as hearsay and inadmissible as evidence.

It must be borne in mind that during about this same period of time two other famous "fossil men" were produced, one of which, it turned out, was based on a pig's tooth and the other, it was finally discovered, was based on a modern ape's jaw! In 1922, a tooth was discovered in western Nebraska which was declared by Henry Fairfield Osborn, one of the most eminent paleontologists of that day, and several other authorities, to combine the characteristics of chimpanzee, *Pithecanthropus*, and Man! As Mark Twain once said, "There is something fascinating about science. One gets such wholesale returns of conjectures out of such a trifling investment of facts" (*Life on the Mississippi*, p. 156).

Osborn and his colleagues could not quite decide whether the original owner of this tooth should be designated as an ape-like man or a man-like ape. He was given the designation *Hesperopithecus haroldcookii* and became known popularly as Nebraska Man. An illustration of what this creature and his contemporaries supposedly looked like was published in the *Illustrated London News*.[33] In this illustration, *Hesperopithecus* looks remarkably similar to modern man, although brutish in appearance. In 1927, after further collecting and studies had been carried out, it was decided that *Hesperopithecus* was neither a man-like ape nor an ape-like man, but was an extinct peccary, or pig![34] I believe this is a case in which a scientist made a man out of a pig and the pig made a monkey out of the scientist!

In 1912, Arthur Smith Woodward, Director of the Natural History Museum of London, and Charles Dawson, a medical doctor and amateur paleontologist, announced the discovery of a mandible and part of a skull. Dawson had recovered these specimens from a gravel pit near Piltdown, England. The jawbone appeared very simian-like except for the teeth, which seemed to show the type of wear expected for humans rather than that for apes. The skull, on the other hand, appeared to be very man-like.

These two specimens were combined and designated *Eanthropus dawsoni*, "Dawn Man." He became known popularly as Piltdown Man. He was judged to be about 500,000 years old. Although a few experts, such as Boule and Henry Fairfield Osborn, objected to the association of this very ape-like jaw with a human-like skull, the consensus of the world's greatest authorities was that Piltdown Man was indeed an authentic link in the evolution of man.

By 1950 a method had become available for assigning a relative age to fossil bones. This method is dependent on the amount of flouride absorbed by bones from the soil. When the Piltdown bones were subjected to this test, it was discovered that the jawbone contained practically no fluoride and thus was no fossil at all. It was judged to be no older than about the year it was found. The skull did have a significant amount of fluoride, but was estimated to be a few thousand years old rather than 500,000 years old.

With this information at hand, the bones were subjected to a thorough and critical examination. It was discovered that the bones had been treated with iron salts to make them look old, and scratch marks were detected on the teeth, indicating that they had been filed. In other words, Piltdown Man was a complete fraud! A modern ape's jaw and a human skull had been doctored to resemble an ape-man, and the forgery had succeeded in fooling most of the world's greatest experts. The success of this monumental hoax served to demonstrate that scientists, just like everyone else, are very prone to find what they are looking for whether it is there or not. The success of the Piltdown hoax for nearly 50 years in spite of the scrutiny of the world's greatest authorities, along with other stories nearly as dubious, led Lord Zuckerman to declare that it is doubtful if there is any science at all in the search for man's fossil ancestry. His review of the Piltdown hoax is especially revealing.[35]

With these lessons in mind, let us now return to our consideration of Peking Man. We will examine first the evaluation of the fossil remains produced by evolutionists and then we will consider that given by a creationist. For the evolutionist viewpoint we will use the publication *Fossil Men*, an English translation of *Les Hommes Fossiles* by Marcellin Boule and H. M. Vallois and which has been referred to earlier.[30] Boule and Vallois devote an extensive section (pp. 130-146 of the English translation) to *Sinanthropus*, or Peking Man.

The first evidence related to *Sinanthropus* was discovered in 1921, when two molar teeth were recovered from a pocket of bone-remains at the site near the village of Choukoutien. A third molar was found in 1927 and given to Dr. Davidson Black. As related earlier, it was on the basis of this tooth that *Sinanthropus pekinensis* was erected. In 1928, the Chinese paleontologist in charge of the excavations, Dr. W. C. Pei, recovered fragments of crania, two pieces of lower jaw and numerous teeth, which were immediately described in a publication by Black. In 1929, Pei unearthed a well-preserved skullcap resembling that of *Pithecanthropus*. Since that time the site was systematically explored under the supervision of the Geological Survey of China. Eventually the collection described at the beginning of this section was recovered.

It is claimed that a vast cavern once existed in the face of the limestone cliff since the "cave filling" appears on the surface along a distance of 100 yards and is about 150 feet thick. It is said that the roof of the cavern collapsed, burying the old cave-filling.

The *Sinanthropus* fragments are found at many different levels of the fillings. The fossil fauna (the bones of about 100 different animals were found) did not vary from the top to the bottom of the deposit and the *Sinanthropus* remains found at the various levels had everywhere the same characteristics. If these remains were actually found in a real cave-filling as alleged, then this means that during all the time it would have taken to lay down 150 feet of filling, no change in *Sinanthropus* or the animals of the area took place.

All of the skulls are damaged and lack their lower jaw. After the discovery of the skulls described earlier, three other skulls were reported to have been recovered in 1936 while Dr. Franz Weidenreich, an American paleontologist of German extraction, was in charge.

Skull III, actually the first to be discovered, is described in some detail by Boule and Vallois (Boule had visited Peking and Choukoutien and had examined the originals). It was ascribed by Black to an adolescent and by Weidenreich to an individual of eight or nine years of age. Boule and Vallois say that viewed from the top and sides, it bears a striking resemblance to *Pithecanthropus,* and that Skull II in its general contour was even more like that of *Pithecanthropus*. They conclude that "In its totality, the structure of the *Sinanthropus* skull is still very ape-like" (p. 136). A bit later they report that the three crania from Locus L (found in 1936) present the same characters as the skulls just mentioned, but in more accentuated form. In a 1937 publication, Boule referred to the *Sinanthropus* skulls as "monkey-like."[36]

The cranial capacity of the skulls, although admittedly very approximate, was estimated to be about 900 c.c. for the skulls discovered earlier and up to about 1200 c.c. for two of the skulls found in 1936. Boule and Vallois point out that these values are about midway between the higher apes and man.

The features of the lower jaw described by Boule and Vallois were all ape-like except for the shape of the dental arcade (curve of the jaw), which was parabolic as in man, apparently, rather than U-shaped as in the apes. Likewise, all of the features for the teeth given by these authorities were apelike except that no diastema (space) separates the canine teeth from neighboring incisors, as is the case in some species of apes (but not all). Furthermore, though the upper canine teeth were "particularly large," rising appreciably above the level of the other teeth as in the apes and being described as "small tusks," the lower canines look rather like large incisors. Thus, with very few exceptions, the structural features of the jaw and teeth were apelike, but the presence of these few exceptions led Boule and Vallois to state that the mandibles and teeth of *Sinanthropus* denote a large primate more closely allied to man than any other known great ape.

After comparing a table of measurements of *Sinanthropus* to one for *Pithecanthropus*, Boule and Vallois say that the differences are less than those within a single species (namely, Neanderthal Man). They therefore insist at the very least that these two creatures be included within a single genus, although they are willing to permit the species differentiation to stand. Since *Pithecanthropus* enjoys priority, they would assign the name *Pithecanthropus pekinensis* to the Choukoutien creature. Since, as noted earlier in their discussion of *Pithecanthropus*, these authorities had said that, based on the skull and teeth alone, we are dealing with beings, if not identical with, then closely allied to the anthropoids, we wonder if by this close linkage of *Sinanthropus* with *Pithecanthropus*, Boule and Vallois wish to downgrade *Sinanthropus* to a creature, if not identical with, at least closely allied to the anthropoids, or if they wish to upgrade *Pithecanthropus*. Today most evolutionists have elevated *Pithecanthropus* and place *Pithecanthropus* and *Sinanthropus* in a single species, *Homo erectus*.

In their discussion of the relationship of *Sinanthropus* to *Pithecanthropus* (p. 141), Boule and Vallois almost accuse Black of fraud, and at the very least accuse him of total lack of objectivity and of twisting facts. Specifically, they say:

> Black, who had felt justified in forging the term *Sinanthropus* to designate *one* tooth, was naturally concerned to legitimize this creation when he had to describe a skull cap. While acknowledging the great resemblance of this piece to the Javan counterpart, he stressed the differences and demonstrated them by numerical data. Now, on studying his tables of measurements, it is quite evident that the differences observed between *Pithecanthropus* on the one hand, and the various fragments of *Sinanthropus* on the other, far from possessing generic value, are less than the variations recorded within the very natural specific group of *Homo neanderthalensis*.

In other words, since Black had stuck his neck out on the basis of a single tooth (remember "Nebraska Man"!) and had erected the *Sinanthropus* category around that tooth, he felt compelled to color the facts to fit his scheme. What confidence can we have, therefore, in any of the descriptions or models of *Sinanthropus* from the hand of Dr. Black?

A section entitled "A New Discussion of the Facts" appears at the end of the chapter devoted to the discussion of *Sinanthropus* by Boule and Vallois. It is based mainly on a model of

Sinanthropus constructed by Weidenreich (Fig. 14), allegedly on the basis of the material found in 1936. This model is so glaringly different from the earlier descriptions of *Sinanthropus* and from a model of *Pithecanthropus* fashioned by Weidenreich himself (see p. 124 of *Fossil Men*), that I strongly suspect that Weidenreich was guilty of the same lack of objectivity and preconceived ideas that motivated Black. The account by Boule and Vallois in this section also varies so decidedly from earlier descriptions of *Sinanthropus*, published elsewhere by Boule, that it is probable that this section was written by Vallois after the death of Boule (the 1952 edition of *Les Hommes Fossiles* was published after Boule's death and was a revision by Vallois of an earlier edition of this book authored solely by Boule).

Davidson Black died in 1934 and was replaced by Franz Weidenreich. Dr. Pei continued to be in charge of the excavations, and it was his duty to submit his finds to Weidenreich for evaluation. It is reported that he found portions of three skulls in 1936. It was these three skulls (referred to by Boule and Vallois as those from Locus L) on which Weidenreich supposedly based his model.

In the section "A New Discussion of the Facts" no new data is introduced but the reader is asked to examine three photographs by Weidenreich showing several views of three skulls or models: a skull of a female gorilla; Weidenreich's model of his female *Sinanthropus*; and the skull of a Northern Chinese. The reader is then invited to verify for himself that *Sinanthropus* occupies a position intermediate between the Anthropoid Apes and Man. If one accepts uncritically Weidenreich's model of *Sinanthropus* as a true portrayal of the real *Sinanthropus*, then he could hardly reject the above appraisal. As a matter of fact, on the basis of this model some have been led to believe that *Sinanthropus* should not be considered as near-man but should be judged to be fully human.

It should be emphasized that in these photographs, skulls of a gorilla and of a man are being compared to a *model* of the *Sinanthropus* skull fashioned by Weidenreich. When a complete skull is available, the specimen is completely reliable, especially if no distortions took place since burial. Almost always the remains of a skull are fragmentary. In this case, the paleontologist attempts to reconstruct the skull based on the fragments, using filler material to fill in missing fragments and modeling missing parts. The reconstruction is more or less reliable depending on how fragmentary the fossil remains are and on the objectivity of the paleontologist. Models are casts of reconstructions or are fashioned according to what the investigator feels the skull should look like.

Today we have no skulls or fragments of *Sinanthropus* (except two teeth), no reconstruction. All we have available are the *models* fashioned by Weidenreich. How reliable are these models? Are they accurate casts of the originals, or do they reflect what Weidenreich *thought* they should look like? Why do his models differ so greatly from the earlier descriptions? I consider these models of Weidenreich to be totally inadmissible as evidence related to the taxonomic affinities of *Sinanthropus*. If such a case were ever brought to court there would not be the slightest doubt that such hearsay evidence would be ruled inadmissible.

Finally, a most peculiar characteristic of the *Sinanthropus* remains is discussed by Boule and

Vallois. As they put it (p. 145):

> How are we to explain the almost complete absence of long bones and this kind of selection of bony parts all belonging to the skull, in which lower jaws predominate? Weidenreich believed that these selected parts did not come into the cave by natural means, but that they must have been brought there by hunters who chiefly attacked young individuals and chose for preference, as spoils or trophies, heads or parts of heads. In itself, this explanation is thoroughly plausible. *But the problem is to name the hunter.* (Emphasis added.)

All authorities agree that every one of the *Sinanthropus* individuals had been killed by hunters and eaten. All of the skulls had been bashed in near the base in order that the brains might be extracted and eaten. Practically nothing of these creatures was found except fragments of the skull, and that in spite of the fact that fragments of almost 40 different individuals were recovered. The only question remaining unanswered in respect to these circumstances was, *who was the hunter*?

Weindenreich, as do almost all other evolutionists, concludes that the hunter must be *Sinanthropus* himself! He was both the hunted and the hunter! This hypothesis is necessary to preserve the status of *Sinanthropus* as the evolutionary ancestor of man.

Boule and Vallois express serious doubts as to the validity of this theory. They say:

> To this hypothesis, other writers preferred the following, which seemed to them more in conformity with our whole

> body of knowledge: the hunter was a true Man, whose stone industry has been found and who preyed upon *Sinanthropus*

Later on they say:

> We may therefore ask ourselves whether it is not overbold to consider *Sinanthropus* the monarch of Choukoutien, when he appears in the deposit only in the guise of a mere hunter's prey, on a par with the animals by which he is accompanied.

In an article published in 1937 in *L'Anthropologie* (p. 21), Boule wrote:

> To this fantastic hypothesis [of Abbe Breuil and Fr. Teilhard de Chardin], that the owners of the monkey-like skulls were the authors of the large-scale industry, I take the liberty of preferring an opinion more in conformity with the conclusions from my studies, which is that the hunter (who battered the skulls) was a real man and that the cut stones, etc., were his handiwork [the nature of this stone industry will be discussed later].

There is thus very good evidence, "more in conformity with our whole body of knowledge," that the *Sinanthropus* creatures were the victims of hunters who were true Men. If this is so, then *Sinanthropus* could not have been the evolutionary ancestor of man, but must have been a large monkey-like or ape-like creature.

In more recent times Louis Leakey has reported findings that are astounding to evolutionists and which suggest additional difficulties for the evolutionary theory of man's origin. Leakey has found evidence for the contemporary existence of

Australopithecus and *Homo erectus* (the same as Java Man and Peking Man) in Bed II of the Olduvai Gorge.[37, 38] Even more startling is Leakey's report that he had found evidence for the existence of a circular stone habitation hut at the bottom of Bed I![38, 39] Deliberate manufacture of shelters has long been presumed to be attributable only to *Homo sapiens.*

If *Australopithecus* and *Homo erectus* existed contemporaneously, then how could one have been the evolutionary ancestor of the other? And how could either be ancestral to man, when man's artifacts are found at a lower stratigraphic level directly underneath the fossil remains of these creatures? These are very hard questions, indeed. The evolutionary hypothesis for the origin of man becomes less and less plausible as more and more evidence becomes available.

We will now consider the evaluation of *Sinanthropus* by a creationist, the Roman Catholic priest, Rev. Patrick O'Connell. To pit the evaluation of a priest against those of eminent evolutionary paleontologists seems akin to pitting David against Goliath. But perhaps in this case, also, David has found a weak spot in the Goliath.

O'Connell was in China during all of the time the excavations at Choukoutien were being carried out, including the Japanese occupation and for several years after their departure. Although he did not make an on-site investigation, O'Connell had the advantage of seeing the accounts published in China in both Chinese and foreign languages. He became convinced that the public had not been given all the facts and that no "missing link" had been found at Choukoutien. He published his conclusions in his book *Science of Today and the*

Problems of Genesis.[40]

O'Connell believed that the disappearance of the *Sinanthropus* remains was by design rather than an accident of war. The Japanese did not interfere with the work at Choukoutien, and Weidenreich and Pei continued the excavations until Weidenreich left in 1940. O'Connell believes that Pei may have destroyed the fossils before the Chinese government returned to Peking in order to conceal the fact that the models did not correspond to the fossils.

In an article published in the Peking periodical *China Reconstructs* in 1954, Dr. Pei says that the material from Choukoutien was then on display. This included the *casts* or *models* of a few of the skulls of *Sinanthropus* (made by Black and Weidenreich), *fossil remains* of various animals, and a selection of the stone instruments found. It appears then that of the material related to *Sinanthropus*, only the fossil remains of *Sinanthropus* are missing.

The almost universally accepted version of the Choukoutien setting is that the fossils of *Sinanthropus* were found in the cave-filling of a huge cavern, the roof of which had collapsed. The human fossils found at the same site at an upper level supposedly were recovered from an upper cave. There seems to be little evidence that a cave existed at either level. As noted earlier, a huge cavern must be postulated for the lower cave, since the "cave filling" extended along the surface for at least 100 yards. The "upper cave" would have had to be as large or larger, since the debris was scattered over even a larger area. Weidenreich never claimed that a cave existed at the upper level, but referred to it as the "so-called upper cave."

According to O'Connell's reconstruction of the events of Choukoutien, a large-scale industry of quarrying limestone had been carried out there in ancient times. That lime-kilns had been constructed and operated there is indicated by the fact that thousands of quartz stones brought from a distance (no quartz is found at Choukoutien) were found in the debris at both levels. The stones had a layer of soot on one side. Enormous heaps of ashes were found at both levels.

The quarrying had been carried on at the two levels on a front of about 200 yards and to a depth of about 50 yards into the hill. The limestone hill was undermined and collapsed, burying everything at both levels under thousands of tons of stone. It was in these heaps of buried ashes and debris that the skulls of *Sinanthropus* were found.

Stones brought from a distance and dressed for building found beside a limestone quarry and enormous heaps of ashes can only mean one thing according to O'Connell: lime-burning was being carried out. Furthermore, lime production on the scale carried out at Choukoutien must mean houses were being built on a considerable scale.

Regardless of whether O'Connell is right concerning a lime-burning industry at Choukoutien, no other explanation has been given for the extensive stone industry found there. H. Breuil, an authority on the Old Stone Age, was invited to Choukoutien. His report, published in the March, 1932, issue of *L'Anthropologie*, tells us that in a section on the lower level of 132 square meters, 12 meters deep, 2,000 roughly shaped stones were found at the bottom of a heap of ashes and debris which contained the skulls of *Sinanthropus* and the bones of about 100 different animals.

The nature of the tools found at the site according to Breuil was not primitive. The gravers and scrapers and other tools, sometimes of fine workmanship, had many features not found in France until the Upper Paleolithic.[41] This evidence could hardly be used, therefore, as an argument for a great antiquity for *Sinanthropus*.

O'Connell points out that very little attention has been paid to the fact that fossil remains of ten human individuals of modern type were found at an upper level of the same exact site where the skulls of *Sinanthropus* were found. Some books, for example Romer's *Man and the Vertebrates*, make no mention of the fact. Others make no mention of this in the section on *Sinanthropus*, but place this information elsewhere. O'Connell believes that these individuals were killed by a landslide caused by undermining of the limestone cliff during quarrying operations and that this same landslide buried the skulls of *Sinanthropus*. Their bones made up the usual assortment expected for such remains.

An examination of a diagram of the site from which the *Sinanthropus* skulls were recovered (p. 132 of *Fossil Men*) tends to support O'Connell. The disposition of the remains, especially those found in the "vertical offshoot of the main pocket" does not seem to correspond to that expected for a cave-filling.

O'Connell points out that some of the early descriptions of *Sinanthropus* by certain investigators differ quite significantly from the later descriptions and models of Black and Weidenreich. He quotes Teilhard de Chardin as saying (*L'Anthropologie*, 1931) that "*Sinanthropus* manifestly resembles the great apes closely." We have already cited Boule's article in

L'Anthropologie (1937, p. 21) in which he refers to the skulls of *Sinanthropus* as "monkey-like."

There seems to be a progression through the two descriptions of *Sinanthropus* by Black and the third description by Weidenreich based on the skulls found in 1936 (see model shown in Fig. 14) during which *Sinanthropus* becomes more and more man-like. Perhaps this is the only evolution involved in this whole affair!

O'Connell concludes that *Sinanthropus* consisted of the skulls of either large macaques (large monkeys) or large baboons killed and eaten by workers at an ancient quarry. There does seem to be considerable evidence that a lime-burning site was buried under rock and debris at Choukoutien. Whether the creatures whose skulls were discovered there were macaques or baboons, they were monkey-like according to Boule. Finally, Boule and others had leaned to the belief that *Sinanthropus* had been killed and eaten by true Man.

O'Connell terms the representation of *Sinanthropus* as a near-man an outright fraud. We believe at the very least a combination of prejudice, preconceived ideas, and a zeal for fame have been responsible for elevating a monkey-like creature to the status of an ape-like man. The same combination that produced Peking Man also produced Nebraska Man from a pig's tooth, Piltdown Man from a modern ape's jaw, and Leakey's East Africa Man from an australopithecine.

NEANDERTHAL MAN

Neanderthal Man was first discovered over a century ago in a cave in the Neander Valley near Dusseldorf, Germany. He was first classified as *Homo neanderthalensis* and portrayed as a semierect brutish subhuman. This misconception of Neanderthal Man was most likely due to the bias of evolution-minded paleoanthropologists plus the fact that the individual on whom this assessment was made had been crippled with arthritis. Furthermore, it is known that these people suffered severely from rickets, caused by a deficiency of Vitamin D. This condition results in softening of bone and consequent malformation. It is now known that Neanderthal Man was fully erect and in most details was indistinguishable from modern man, his cranial capacity even exceeding that of modern man. It is said that if he were dressed in a business suit, and were to walk down one of our city streets, he would be given no more attention than any other individual. Today he is classified *Homo sapiens* — fully human. It is claimed that he lived as long as about 100,000 years ago to as recently as about 25,000 years ago.

Other undoubted human remains, such as Swanscombe Man, have been dated at almost 250,000 years.[13] These dates, and those assigned to other fossils, such as Peking Man, are only estimates based on assumptions made by evolutionary geologists. They assume that the Pleistocene Epoch, in which all of these fossils are said to occur, beginning with *Australopithecus*, began about two or three million years ago. They had earlier estimated the duration of the Pleistocene to be only a fraction of that, but stretching out the Pleistocene has given evolutionists the time they believe is required for the australopithecines to evolve into men.

If a fossil is believed to have been found in the

early Pleistocene, it is assigned an age of 2-3 million years. If it is estimated to be of middle Pleistocene age, it would be estimated to be about 1.5 million years old. It would be assigned a younger age if believed to be of late Pleistocene times. The finer points of this dating system are described in the book by Day.[42] The validity of any such assigned date depends on many things, but mainly on the assumption concerning the duration of the Pleistocene.

ALL HUMAN EVOLUTION SCHEMES SHATTERED IF RICHARD LEAKEY IS CORRECT

The latest reports of Richard Leakey[43, 44] are startling, and, if verified, will reduce to a shamble the presently held schemes of evolutionists concerning man's origin. As described above, man supposedly descended through an ape-like ancestor, *Australopithecus* (Dr. Louis Leakey's *Zinjanthropus*), allegedly about 2 million years ago, and then later through a near-man ancestor, Java Man — Peking Man, claimed to have been about one-half million years old. Now Richard Leakey reports that his team has discovered a skull even more modern than so-called Peking Man in a deposit supposedly dating nearly 3 million years old!

The skull, found near Lake Turkana (previously Lake Rudolph) in Kenya, East Africa, does not have the heavy brow ridges of Peking Man, and the wall of the skull is thin as in modern man. When the pieces were all fitted together, the shape of the skull was similar to modern man, apparently almost indistinguishable from those of many individuals living today.

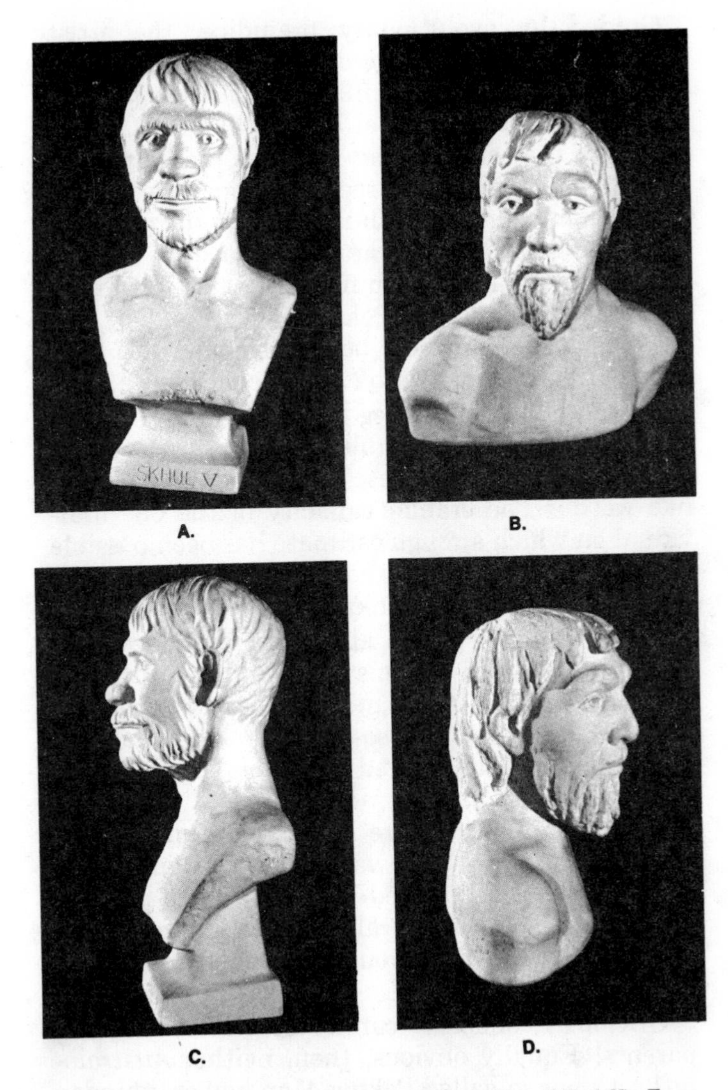

FIGURE 15. Two flesh models of Neanderthal Man (Skhul V). From Rusch's *Human Fossils*, in *Rock Strata and the Bible Record*, P. A. Zimmerman, Ed., Concordia Publishing House.

Guided by evolutionary prejudice, the artist chosen by *National Geographic* to portray a flesh reconstruction of the skull gave it a simian-like nose rather than a typical human-like nose.[44] The nature of the soft anatomy, such as the hair, ears, and nose, cannot be discerned from osteological remains, of course, and thus these are the products of the imagination of the artist. Replace the simian-like nose with a typical human-like nose, as in the flesh reconstruction of Peking Man (Fig. 14), and the reconstruction would be indistinguishable from types of individuals we have all seen many times.

According to reports, leg bones found in the same strata are indistinguishable from those of modern man, indicating that these individuals walked just like we do. The cranial capacity of the one individual on which a rough estimate has been possible is surprisingly small — about 800 c.c. The age and sex of this creature is unknown, however.

If the results reported above are verified, and if the ages assigned to this find and to *Australopithecus* and Peking Man are accepted, then the australopithecines and so-called Peking Man are demolished as our alleged ancestors. This find of Richard Leakey's team, according to Leakey and his collaborators, is more modern in almost every respect, and yet it was contemporary with *Australopithecus*, alleged to be our ape-like ancestor, and supposedly almost 2½ million years older than Peking Man, our assumed near-man ancestor.

Obviously, offspring cannot be older than their parents! Equally obvious, then, neither *Australopithecus* nor so-called Peking Man had anything to do with the origin of man.

These results certainly support the contention,

expressed earlier in this chapter, that *Australopithecus*, while possessing canines and incisors more similar to man than those of extant apes, was nonetheless nothing but an ape. After all, teeth alone don't make the man! These results also support the conclusions expressed earlier about Peking Man.

Early in 1973, Richard Leakey gave a lecture in San Diego describing his latest results. He stated his convictions that these findings simply eliminate everything we have been taught about human origins and, he went on to say, he had nothing to offer in its place! Creationists *do* have something to offer in its place, of course. We believe that these results support man's special creation rather than his origin from an animal ancestry. These results also strongly support our belief that man and the ape have always coexisted.

EVOLUTIONIST SUGGESTS MONKEYS AND APES EVOLVED FROM MAN!

Leakey's announcements have predictably generated much controversy as well as scrambling to revise current theories on the evolution of man. Some paleoanthropologists, dismayed by this threat to a long-held and easily taught theory, have declared that either Leakey's evidence, or his evaluation of the evidence, is faulty. Others believe Leakey is correct, and have either declared that they are left without any theory at all or, at least in one case, have suggested radical departures from current theories.

Ever since Darwin, evolutionists have maintained that man has evolved from some ape-like creature. Now Geoffrey Bourne, Director of the

Yerkes Regional Primate Research Center of Emory University, is saying that evolutionists have been wrong all along and that just the opposite is true — monkeys and apes have evolved from man!

An article published in *Modern People* (Vol. 1, p. 11, April 18, 1976) states:

> For, whereas Darwin popularized the theory that man descended from the primate family, Dr. Bourne believes the exact opposite — that monkeys, apes, and all other lower primate species are really the offspring of man.

Bourne cites as evidence for this incredible theory the findings of Richard Leakey and evidence from embryology. The article states:

> Another argument the doctor, who is considered one of the world's leading experts on primates, uses to support his theory, is the fact that an ape fetus looks like a human fetus during the early stages of development before birth. It is only in the later stages of gestation that the unborn ape starts developing typical apelike characteristics.
>
> This means the development of an ape infant recapitulates his origin; he goes from a human-like to an ape-like animal in the fetus.

This latter statement contradicts earlier claims by evolutionists that just the opposite is true, that the human embryo recapitulates its evolutionary history in such a way that the embryo looks more like an ape in its earlier development and looks more human-like as it develops further. Furthermore, the entire idea of embryological recapitulation has been discredited by modern embryologists.

Bourne's theory also contradicts the Darwinian idea of survival of the fittest, or natural selection, as the driving force of evolution, for the article further relates that:

> In speaking about man, the doctor claims that it is man's superior brain, formation of his hands and arms, and his position in walking that have made him the most dominant animal on Earth.

If this is true, apes and monkeys are obviously inferior to man. How, then, did natural selection bring about the evolution of apes and monkeys from man?

Bourne, in a personal communication, confirmed the accuracy of the article. How can it be that evolutionists, looking at exactly the same data, arrive at theories that are diametrically opposed to one another — in one case that man has evolved from an ape-like animal, and in the other that apes have evolved from man? Incredible! Incredible unless the basic assumption on which their entire system of belief is based — evolution — is wrong.

Lord Zuckerman, himself an evolutionist, has, as noted earlier, expressed his conviction that there is really no science at all in the search for man's fossil ancestry. He has arranged various scientific endeavors in a spectrum beginning with what he considered to be pure science and moving toward endeavors he considered to be less and less scientific. He began with chemistry and physics, then moved into the biological sciences and then into the social sciences. He then goes on to say:

> We then move right off the register of objective truth into those fields of presumed biological science, like extrasensory perception or the interpretation of man's fossil history, where to the faithful anything is possible — and where the ardent believer is sometimes able to believe several contradictory things at the same time.[45]

Lord Zuckerman apparently believes that, if man has evolved, he managed to do so without leaving a trace in the fossil record. He states:

> For example, no scientist could logically dispute the proposition that man, without having been involved in any act of special creation, evolved from some ape-like creature in a very short space of time — speaking in geological terms — without leaving any fossil traces of the steps of the transformation.[35]

In other words, if one excludes special creation as a possible explanation for man's origin, man must have evolved without leaving a trace of that evolution in the fossil record according to Lord Zuckerman. Furthermore, Lord Zuckerman's research in this field could, in the finest sense of the word, be called true scientific research. No one more richly deserves the title of an "authority" in his field than does Lord Zuckerman.

If, as Lord Zuckerman has stated, man has appeared on this planet without leaving any fossil traces of a transformation from an ape-like creature, then the actual evidence accords fully with the concept of the direct special creation of man, but is contradictory to a supposed gradual evolutionary transformation through millions of years from some ape-like creature.

WHERE DID CAVE MEN COME FROM?

Where did such "cave men" as Neanderthal, Cro-Magnon, and Swanscombe Men come from? Creationists believe that they were migrants from a more highly developed population center. Neanderthal Man, for example, abruptly appeared in Europe. Anthropologists have no evidence whatever concerning his origin. He as abruptly disappears and is replaced by Cro-Magnon Man whose origin is equally mysterious to evolutionists. Cro-Magnon Man cannot be distinguished from modern Europeans. Neanderthal Man and Cro-Magnon Man were different races, apparently, of our species, *Homo sapiens*.

As members of a species disperse into small groups, such that they become geographically isolated, they become reproductively isolated as well. Each such group will carry with it only a fraction of the total gene pool, or genetic characteristics, of the population from which it split off. Being a small group, a high degree of in-breeding will result. Such a process may result in the rapid surfacing of genetic traits that were previously suppressed in the large population due to dilution through intermarriage throughout the entire population. As a result, "tribes" or "races" arise.

As this dispersal occurred from the original center of population, these small groups may have carried few skills with them or they eventually may have lost some of their original skills. Scattering in small groups may contribute to this loss via several factors. Lack of population pressure results in a reduction of the need for weapons to defend territory and to defend against predatory raids. Weapons may thus be abandoned. Lack of population pressure may also result in the abandonment of the practice of agriculture, since simple food gathering may suffice to feed the group. Furthermore, ideas and skills are no longer interchanged with neighboring groups. "Progress," as we generally understand it, could be severely retarded, and even a "degeneration" into a more "primitive" state might result.

The Tasaday people of the interior of Mindanao, a large southern island of the Philippines, constitute an example of this process in relatively recent times.[46] Although no one really knows, it has been estimated that these people became isolated from fellow Filipinos about 500-1000 years ago. Surely the Filipinos at that time were practicing agriculture and were manufacturing a variety of tools and weapons.

Today, however, the Tasadays, after a long period of isolation, and relieved of the pressures of competing for space, food, and other necessities, possess no knowledge of agriculture and, except for tools given to them, possess only a few very crude stone and bamboo tools and no weapons. They retain little of what we understand as culture. They certainly would be considered primitive.

Thus, while civilization was developing relatively rapidly in the heavily populated portions of Asia and Europe, peoples in the sparsely settled areas of Europe, the Americas, Australia, and southern Africa continued in a relatively primitive state, some even to this day. It is not surprising, therefore, that the remains of fossil man and associated artifacts, scattered as early man was, indicate that he lived in an uncivilized state, although the Neanderthal people actually possessed a much higher culture and level of technology than the Tasadays. They manufactured rather sophisticated tools and weapons of stone. They were a religious people, as evidenced by the burial of their dead with flowers and various objects they believed would be useful in a life to follow.

That evolution theory, in view of known genetic data, produces no satisfactory explanation for the origin of races is evident from the following statement made in 1972 by the famous evolutionist, the late Theodosius Dobzhansky:

> It is almost incredible that a century after Darwin, the problem of the origin of racial differences in the human species remains about as baffling as it was in his time.[47]

In other words, there is no way of correlating the genetic data associated with the various races within an evolutionary framework. It is an amazing thing that evolutionists insist that they can explain how the universe evolved, how life evolved, how fishes, amphibians, reptiles, birds, and mammals evolved, how primates evolved from earlier mammals, and how apes, monkeys, and men evolved from earlier primates, and yet they must admit that they cannot explain the origin of races within the species *Homo sapiens*! If evolution theory cannot even explain the origin of races in the light of the known scientific evidence, how then can one pretend to use this theory to explain the most profound mysteries of all? Apparently the closer the theory approaches the actual scientific data, the more untenable it becomes.

One obvious racial difference is skin color. It has

sometimes been suggested that the negroid race became black as an adaptation to the more intense ultraviolet light from the sun in the tropics. This idea leaves unanswered why people equally black are not found in other areas of equally intense ultraviolet light, such as South America and North America. Creationists believe that skin color variations developed as a natural sorting out of preexisting genetic traits during the formation of races as described in the section above. According to this view, blacks tended to migrate into those areas where their black skin offered protection from intense sunlight, while the fair-skinned, blue-eyed Scandinavian race naturally migrated to the far north to escape the more intense ultraviolet light encountered near the equator.

SUMMARY

Examination of the record of order Primates gives us the same picture found throughout the remainder of the fossil record. Basically different types, whether lemurs, tarsiers, monkeys, apes, or men, appear suddenly in the fossil record, and transitional forms between these basic types cannot be found. The true situation has been well-stated by Rusch in his excellent review of fossil man:

> Therefore it may be concluded that fossil evidence offers no support for any schemes of the evolutionary descent of man, either within hominid genera or from primate ancestors.[48]

REFERENCES

1. E. L. Simons, *Annals New York Academy of Sciences*, Vol. 167, p. 319 (1969).
2. A. S. Romer, *Vertebrate Paleontology*, 3rd Edition, The University of Chicago Press, Chicago, 1966, p. 218.
3. A. J. Kelso, *Physical Anthropology*, 2nd Edition, J. B. Lippincott, New York, 1974, p. 142.
4. Ref. 2, p. 221.
5. Ref. 3, p. 150.
6. Ref. 3, p. 151.
7. E. L. Simons, *Annals New York Academy of Sciences*, Vol. 102, p. 293 (1962).
8. E. L. Simons, *Scientific American*, Vol. 211, p. 50 (1964).
9. Reference 2, p. 224.
10. D. R. Pilbeam, *Nature*, Vol. 219, p. 1335 (1968).
11. D. R. Pilbeam, *Advancement of Science*, Vol. 24, p. 368 (1968).
12. E. L. Simons and D. R. Pilbeam, *Science*, Vol. 173, p. 23 (1971).
13. R. B. Eckhardt, *Scientific American*, Vol. 226, p. 94 (1972).
14. E. L. Simons and D. R. Pilbeam, *Folia Primatol.*, Vol. 3, p. 81 (1965).
15. C. J. Jolly, *Man*, Vol. 5, p. 5 (1970).
16. D. R. Pilbeam, *Nature*, Vol. 225, p. 516 (1970).
17. D. R. Pilbeam, *The Evolution of Man*, Funk and Wagnalls, New York, 1970.
18. R. Broom and G. W. H. Schepers, *Transv. Mus. Mem.*, Vol. 2, pp. 1-272 (1946).

19. W. E. LeGros Clark, *Journal of Anatomy London,* Vol. 81, pp. 300-333 (1947).
20. S. Zuckerman, *Journal of the Royal College of Surgeons of Edinburgh*, Vol. 11, pp. 87-115 (1966).
21. S. Zuckerman, *Beyond the Ivory Tower*, Taplinger Pub. Co., New York, 1970, pp. 75-94.
22. C. Oxnard, *University of Chicago Magazine*, Winter 1974, pp. 8-12.
23. C. Oxnard, *Nature*, Vol. 258, pp. 389-395 (1975).
24. Ref. 21, p. 77.
25. Ref. 23, p. 389.
26. Ref. 22, pp. 11-12.
27. A. Montagu, *Man: His First Million Years*, World Publishers, Yonkers, N.Y., 1957, pp. 51, 52.
28. J. T. Robinson, *Nature*, Vol. 205, p. 121 (1965).
29. W. Howells, *Mankind in the Making*, Doubleday and Co., Garden City, N.Y., 1967, pp. 155-156.
30. M. Boule and H. M. Valois, *Fossil Men*, The Dreyden Press, New York, 1957, p. 118. This is the English translation of the 1952 edition of *Les Hommes Fossiles*.
31. Ref. 30, p. 122.
32. Ref. 30, p. 123.
33. *Illustrated London News*, June 24, 1922.
34. W. K. Gregory, *Science*, Vol. 66, p. 579 (1927).
35. Ref. 21, p. 64.
36. M. Boule, *L'Anthropologie*, 1937, p. 21.
37. M. D. Leakey, *Olduvai Gorge Vol. 3*, Cambridge University Press, Cambridge, 1971, p. 272.

38. A. J. Kelso, *Physical Anthropology*, 1st Ed., J. B. Lippincott Co., New York, 1970, p. 221.
39. Ref. 37, p. 24.
40. P. O'Connell, *Science of Today and the Problems of Genesis. Book 1.* Christian Book Club of America, Hawthorne, CA., 1969.
41. Ref. 30, footnote p. 145.
42. M. Day, *Guide to Fossil Man*, The World Publishing Co., New York, 1965, p. 13.
43. *Science News*, Vol. 102, p. 324 (1972).
44. R. E. Leakey, *National Geographic*, Vol. 143, p. 819 (1973).
45. Ref. 21, p. 19.
46. K. MacLeish, *National Geographic*, Vol. 142, p. 219 (1972).
47. T. Dobzhansky, in *Sexual Selection and the Descent of Man*, B. Campbell, ed., Aldine Pub. Co., Chicago, 1972, p. 75.
48. W. H. Rusch, "Human Fossils," in *Rock Strata and the Bible Record*, ed. P. Zimmerman, Concordia Publishing House, St. Louis, 1970, p. 172.

Chapter VII
THE FOSSILS SAY NO

In the preceding chapters, we cited example after example of failure to find transitional forms where evolutionary theory predicts such forms should have been found. Some might suspect that we have biased our choice of examples in such a way that only those cases have been cited where transitional forms have not yet been found, while failing to mention many other examples where transitional forms between basically different kinds of animals or plants are known. Nothing could be further from the truth.

The examples cited in this book are in no way exceptions but serve to illustrate what is characteristic of the fossil record. While transitions at the subspecies level are observable and some at the species level may be inferred, the absence of transitional forms between higher categories (the created kinds of the creation model) is regular and systematic. We now propose to document this statement by citing published statements of evolutionists.

We wish to cite first the world's foremost evolutionary paleontologist, George Gaylord Simpson. In his book, *Tempo and Mode in Evolution* under the section entitled, "Major Systematic Discontinuities of Record," he states that nowhere in the world is there any trace of a fossil that would close the considerable gap between *Hyrocotheruim*, which most evolutionists assume was the first horse, and its supposed ancestral order Condylarthra. He then goes on to say:

> This is true of *all* the thirty-two orders of mammals. . . .*The earliest and most primitive known members of every order already have the basic ordinal characters,* and in no case is an approximately continuous sequence from one order to another known. In most cases the break is so sharp and the gap so large that the origin of the order is speculative and much disputed.[1] (Emphasis added.)

Later on (p. 107) Simpson states:

> This *regular absence* of transitional forms is not confined to mammals, but is an almost *universal phenomenon*, as has long been noted by paleontologists. It is true of almost all orders of all classes of animals, both vertebrate and invertebrate. A fortiori, it is also true of the classes, and of the major animal phyla, and it is apparently also true of analogous categories of plants. (Emphasis added.)

In his book, *The Meaning of Evolution*, Simpson, with reference to the appearance of new phyla, classes, or other major groups, states that:

> The process by which such *radical events* occur in evolution is the subject of

> one of the most serious remaining disputes among qualified professional students of evolution. The question is whether such *major events* take place *instantaneously*, by some process essentially unlike those involved in lesser or more gradual evolutionary change, or whether all of evolution, including these major changes, is explained by the same principles and processes throughout, their results being greater or less according to the time involved, the relative intensity of selection, and other material variables in any given situation.
>
> Possibility for such dispute exists because transitions between major grades of organization are seldom well recorded by fossils. There is in this respect a tendency toward *systematic deficiency* in the record of the history of life. *It is thus possible to claim that such transitions are not recorded because they did not exist*, that the changes were not by transition but by sudden leaps in evolution.[2] (Emphasis added.)

If phyla, classes, orders, and other major groups were connected by transitional forms rather than appearing suddenly in the fossil record with basic characteristics complete, it would not be necessary, of course, to refer to their appearance in the fossil record as "radical events." Furthermore, it cannot be emphasized too strongly that even evolutionists are arguing among themselves whether these major categories appeared *instantaneously* or not! It is precisely the argument of creationists that these forms *did* arise *instantaneously* and that the transitional forms are not recorded because they never existed! Creationists thus would re-word Simpson's statement to read:

> It is thus possible to claim that such transitions are not recorded because they did not exist, that these major types arose by creation rather than by a process of gradual evolution.

In a more recent work, Simpson stated that, "It is a feature of the known fossil record that most taxa appear abruptly." In the same paragraph he states further that, "Gaps among known species are sporadic and often small. *Gaps among known orders, classes, and phyla are systematic and almost always large.*"[3] (Emphasis added.)

Although we intend to do so, it would hardly be necessary to document further the nature of the fossil record. It seems obvious that if the above statements of Simpson were stripped of all presuppositions and presumed evolutionary mechanisms to leave the bare record, they would describe exactly what is required by the creation model. This record is woefully deficient, however, in the light of the predictions of the evolution model.

No one has devoted himself more wholeheartedly than Simpson to what Dobzhansky[4] has called "the mechanistic materialist philosophy shared by most of the present establishment in the biological sciences." Simpson asserts that most paleontologists "find it logical, if not scientifically required, to assume that the sudden appearance of a new systematic group is not evidence for creation. . . ."[5]

Simpson has thus expended considerable effort in attempts to bend and twist every facet of evolution theory to explain away the deficiencies of the fossil record.[6-8] One needs to be reminded, however, that if evolution is adopted as an *a priori* principle, it is always possible to imagine auxiliary hypotheses — unproved and by nature unprovable — to make it work in any specific case. By this process biological evolution degenerates into what Thorpe calls one of his "four pillars of unwisdom" — mental evolution that is the result of random tries preserved by reinforcements.[9]

In reference to the nature of the record, Arnold has said:

> It has long been hoped that extinct plants will ultimately reveal some of the stages through which existing groups have passed during the course of their development, but it must freely be admitted that this aspiration has been fulfilled to a very slight extent, even though paleobotanical research has been in progress for more than one hundred years.[10]

The following remarks of Professor E. J. H. Corner of the Cambridge University botany school were refreshingly candid:

> Much evidence can be adduced in favor of the theory of evolution — from biology, biogeography, and paleontology, *but I still think that to the unprejudiced, the fossil record of plants is in favor of special creation.*[11] (Emphasis added.)

This evolutionist frankly states that the fossil record of plants does not support evolution, but rather supports creation!

Olson has said:

> A third fundamental aspect of the record is somewhat different. Many new groups of plants and animals suddenly

> appear, apparently without any close ancestors. Most major groups of organisms — phyla, subphyla, and even classes — have appeared in this way . . . . The fossil record which has produced the problem, is not much help in its solution. . . . Most zoologists and the majority of paleontologists feel that the breaks and the abrupt appearances of new groups can be explained by the incompleteness of the record. Some paleontologists disagree and believe that these events tell a story not in accord with the theory and not seen among living organisms.[12]

In regard to the remark concerning the alleged incompleteness of the record, we refer to the statement by George recorded earlier in this book about the richness of the record. Further refutation of that explanation for the discontinuities may be deduced from Newell's statement that, "Many of the discontinuities tend to be more and more emphasized with increased collecting."[13]

In their recent book on the principles of paleontology, Raup and Stanley have remarked:

> Unfortunately, the origins of most higher categories are shrouded in mystery: commonly new higher categories appear abruptly in the fossil record without evidence of transitional forms.[14]

Du Nouy has described the evidence in this way:

> In brief, each group, order, or family seems to be born suddenly and we hardly ever find the forms which link them to the preceding strain. When we discover them they are already completely differentiated. Not only do we find practically

> no transitional forms, but in general it is impossible to authentically connect a new group with an ancient one.[15]

Kuhn has remarked:

> The fact of descent remains. However, descent beyond the typologically circumscribed boundaries is *nowhere demonstrable*. Therefore, we can indeed speak about a descent within types, but *not about a descent of types.*[16] (Emphasis added.)

Concerning the major groups or phyla, Clark has stated:

> No matter how far back we go in the fossil record of previous animal life upon earth, we find no trace of any animal forms which are intermediate between the various major groups or phyla.[17]

Later on in this same volume (p. 196) he says:

> Since we have not the slightest evidence, either among the living or the fossil animals, of any intergrading types following the major groups, it is a fair supposition that there never have been any such intergrading types.

A reviewer of the recent book, *Evolutionary Biology, Volume 6*[18] states that:

> Three paleontologists (no less) conclude that stratigraphic position is totally irrelevant to determination of phylogeny and almost say that no known taxon is derived from any other.[19]

Even in the famous horse "series" which has been so highly touted as proof for evolution within the order, we find that transitional forms between major types are missing. Thus du Nouy has stated in reference to horses:

> But each one of these intermediaries seems to have appeared "suddenly," and it has not yet been possible, because of the lack of fossils, to reconstitute the passage between these intermediaries. Yet it must have existed. The known forms remain separated like the piers of a ruined bridge. We know that the bridge has been built, but only vestiges of the stable props remain. The continuity we surmise may never be established by facts.[20]

Goldschmidt has said, "Moreover, within the slowly evolving series, like the famous horse series, the decisive steps are abrupt without transition."[21] Frank Cousins has recently published an excellent paper exposing many of the weaknesses and fallacies in the use of fossil horses as evidence for evolution.[22]

Goldschmidt, in contrast to Simpson and the majority of evolutionists, accepted the discontinuities in the fossil record at face value. He rejected the neo-Darwinian interpretation of evolution (the modern synthesis in today's terms), which is accepted by almost all evolutionists, at least among those who accept any theory concerning mechanisms at all. The neo-Darwinian interpretation supposes that all evolutionary changes took place slowly and gradually via many thousands of slight changes. Goldschmidt instead proposed that major categories (phyla, classes, order, families) arose instantaneously by major saltations or sytemic mutations.[23, 24]

Goldschmidt termed his mechanism the "hopeful monster" mechanism. He proposed, for instance, that at one time a reptile laid an egg and a bird was hatched from the egg! All major gaps in the fossil record were accounted for, according to Goldschmidt, by similar events — something laid an egg, and something else got born! Neo-Darwinists prefer to believe that Goldschmidt is the one who laid the egg, maintaining that there is not a shred of evidence to support his "hopeful monster" mechanism. Goldschmidt insists just as strongly that there is no evidence for the postulated neo-Darwinian mechanism (major transformations by the accumulation of micromutations). Creationists agree with both the neo-Darwinists and Goldschmidt — there is no evidence for *either* type of evolution! Goldschmidt's publications do offer cogent arguments against the neo-Darwinian view of evolution, from both the field of genetics and the field of paleontology.

No one was more wholly committed to evolutionary philosophy than was Goldschmidt. If anybody wanted to find transitional forms, he did. If anybody would have admitted that a transitional form was a transitional form, if indeed that's what it was, he would have. But concerning the fossil record, this is what Goldschmidt had to say:

> The facts of greatest general importance are the following. When a new phylum, class, or order appears, there follows a quick, explosive (in terms of geological time) diversification *so that practically all orders or families known appear suddenly and without any apparent transitions.*[21] (Emphasis added.)

Now, creationists ask, *what better description of the fossil record could one expect, based on the predictions of the creation model*? On the other hand, unless one accepts Goldschmidt's "hopeful monster" mechanism of evolution, this description contradicts the most critical prediction of the evolution model — the presence in the fossil record of the intermediates demanded by the theory.

Some critics might complain that the publications by Goldschmidt espousing the "hopeful monster" mechanism are from 25 to 40 years old, and, furthermore, his ideas have been discredited by modern evolutionists. The important point is, however, why did Goldschmidt feel forced to propose such an incredible mechanism in the first place? Goldschmidt felt forced to propose this mechanism because transitional forms between kinds cannot be found, each kind appearing in the fossil record fully formed. Intense searching of the fossil record during the past quarter century has produced nothing that would have caused Goldschmidt to change his mind.

Furthermore, just recently one of America's best known evolutionists has rallied to the defense of Goldschmidt's ideas. Stephen Jay Gould, a professor at Harvard University teaching geology, biology, and the history of science, among his many publishing activities writes articles which appear in each issue of the bi-monthly journal, *Natural History*, a publication of the American Museum of Natural History. In a recent issue, he published an article entitled "The Return of the Hopeful Monsters." [25]

After recounting the "official rebuke and derision" poured out on Goldschmidt by his fellow evolutionists because of his "hopeful monster" mechanism, Gould says: "I do, however, predict that during the next decade Goldschmidt will be largely vindicated in the world of evolutionary biology." A little later he states: "The fossil record with its abrupt transitions offers no support for gradual change . . . ."

Somewhat later in this same article Gould says:

> All paleontologists know that the fossil record contains precious little in the way of intermediate forms; transitions between major groups are characteristically abrupt.

Gould is thus arguing that the fossil record, just as Goldschmidt argued, does not produce evidence of the gradual change of one plant or animal form into another and that, again, just as Goldschmidt argued, each kind appeared abruptly.

Gould then introduces another argument against gradual change that was used by Goldschmidt. Gould says:

> Even though we have no direct evidence for smooth transitions, can we invent a reasonable sequence of intermediate forms, that is, viable, functioning organisms, between ancestors and descendants? Of what possible use are the imperfect incipient stages of useful structures? What good is half a jaw or half a wing?

The argument here, that gradual evolutionary change of one form into another is impossible because the transitional forms, being incomplete, could not function, is an argument that has long been suggested by creationists. This was one of Goldschmidt's key arguments against the neo-Darwinian mechanism of evolution and is now being echoed by Gould.

Gould argues, as did Goldschmidt, that most large evolutionary changes are brought about by small alterations in rates of development. In the first place, there is not one shred of empirical evidence to support such an idea. Even Goldschmidt admitted that no one had ever seen anything like that happen (that is, a new type arising by the postulated hopeful monster mechanism). Gould, in the article referred to above, cited Goldschmidt's work that allegedly showed that large differences in color pattern in caterpillars resulted from small changes in the timing of development. Of course, to cite this as evidence in support of the hopeful monster mechanism is sheer nonsense. The only change produced was in the color of the caterpillar. It remained the same species, as did the butterfly that was produced from the caterpillar. Are such processes supposed to explain the origin of the caterpillar and butterfly in the first place? Of course not. In fact the clear thrust of Gould's article is that major evolutionary changes were *not* produced by such minor variations. Variations in the color patterns of caterpillars then offers no support whatsoever for the idea that major evolutionary changes occur through small variations in rates of development.

According to Goldschmidt, and now apparently according to Gould, a reptile laid an egg from which the first bird, feathers and all, was produced. How, one may ask, were entirely new and novel structures, such as feathers, produced all at once by small variations in rates of development of entirely different structures? A feather is an amazingly complex structure with many elements marvelously designed to function together in such a way that the feather performs its task in an optimal fashion. Its very existence speaks of deliberate design. To believe that a feather, or an eye, or a kidney, let alone a new plant or animal, could be produced from an animal that possessed none of these by small variations in rates of development is absolutely incredible.

But, according to Gould, this appears to be what evolutionists must believe. On the final page of the article cited above, Gould says:

> Indeed, if we do not invoke discontinuous change by small alteration in rates of development, I do not see how most major evolutionary transitions can be accomplished at all. Few systems are more resistant to basic change than the strongly differentiated, highly specified, complex adults of "higher" animal groups. How could we ever convert a rhinoceros or a mosquito into something fundamentally different. Yet transitions between major groups must have occurred in the history of life.

It seems evident that evolutionists will believe anything, no matter how incredible, in order to salvage their fundamental dogma that "transitions between major groups must have occurred in the history of life." Indeed, few systems are more resistant to basic change than the strongly differentiated, highly specified, complex adults of higher animal groups. To believe that there exists a phylogenetic tree whereby a fish was converted into an amphibian, an amphibian into a reptile, a reptile into a "lower" mammal, a "lower" mammal into a "lower" primate, a "lower" primate into an ape, and an ape into a man all by discontinuous changes due to small alterations in rates of development is incredibly astounding, especially when such an idea is suggested by a highly intelligent, well-trained scientist.

It seems evident that if such a well-established modern-day evolutionist as Stephen Jay Gould feels forced to postulate that evolution has occurred by the hopeful monster mechanism, then indeed there is no evidence that evolution is occurring in the present as postulated by neo-Darwinists. If there were such evidence, no one would feel forced to adopt the incredible hopeful monster mechanism. But, on the other hand, certainly no one has ever witnessed the birth of a hopeful monster. Nor evidently is there any evidence that evolution has occurred in the past, since the only such evidence would be the existence of transitional forms, the absence of which the hopeful monster mechanism was invented to explain! It seems to me that the "return of the hopeful monsters," rather than explaining the absence of skeletons in the museums, will only produce a few more in the closets of evolutionists to be explained.

Gould has made a number of revealing statements in other articles in *Natural History*. He has said, for example:

> The extreme rarity of transitional forms in the fossil record persists as the trade secret of paleontology. The evolutionary trees that adorn our textbooks have data only at the tips and nodes of their branches; the rest is inference, however reasonable, not the evidence of fossils.[26]

Later in the same article he states:

> The history of most fossil species includes two features inconsistent with gradualism: 1. *Stasis*. Most species exhibit no directional change during their tenure on earth. They appear in the fossil record looking much the same as when they disappear; morphological change is usually limited and directionless. 2. *Sudden Appearance*. In any local area, a species does not arise gradually by the steady transformation of its ancestors; it appears all at once and "fully formed."

In an article discussing taxonomical classifications, Gould said:

> The three-leveled, five-kingdom system may appear, at first glance, to record an inevitable progress in the history of life that I have often opposed in these columns. Increasing diversity and multiple transitions seem to reflect a determined and inexorable progression toward higher things. But the paleontological record supports no such interpretation. There has been no steady progress in the higher development of organic design. We have had, instead, vast stretches of little or no change and one evolutionary burst that created the whole system.[27]

Eliminate the words "evolutionary burst" and substitute the words "burst of creation" and one would think he was reading an article by a creationist.

In a recent article, David B. Kitts, professor in the Department of Geology at the University of Oklahoma and an evolutionist who received his training in vertebrate paleontology under George Gaylord Simpson, said:

> Despite the bright promise that paleontology provides a means of "seeing" evolution, it has presented some nasty difficulties for evolutionists, the most notorious of which is the presence of "gaps" in the fossil record. Evolution requires intermediate forms between species and paleontology does not provide them. . . .[28]

Macbeth says flatly:

> *Darwinism has failed in practice.* The whole aim and purpose of Darwinism is to show how modern forms descended from ancient forms, that is, to construct reliable phylogenies (genealogies or family trees). In this it has utterly failed.[29] (Emphasis added.)

He then goes on to quote other authors to the effect that the phylogenies found in textbooks are based on unsupported assertions, imaginative literature, speculations, and little more.

CREATION, EVOLUTION, AND THE FOSSIL RECORD: SUMMARY

The major predictions of the creation model were:

1. The abrupt appearance of highly complex and diverse forms of life with no evidence of ancestral forms.
2. The sudden appearance of basic plant and animal kinds without evidence of transitional forms between these basic kinds.

The fossil record reveals:

1. The abrupt appearance of a great variety of highly complex forms of life. No evolutionary ancestors for these animals can be found anywhere on the earth.
2. The sudden appearance of the higher categories of plants and animals with no evidence of transitional forms between these basic kinds.

The historical, or fossil, record thus provides excellent support for special creation, but contradicts the major predictions of evolution theory. In answer to the question, did evolution really occur, the fossils shout a resounding NO!

EMBRYOLOGY, VESTIGIAL ORGANS, AND HOMOLOGY

What about other evidence for evolution, such as that from embryology, homology, and vestigial organs? Almost all evolutonists used to believe (and many still do) that the human embryo (and all other embryos), during its development, takes on, successively, the appearance of its evolutionary ancestors in the proper evolutionary sequence. Ontogeny (embryological development) is said to recapitulate phylogeny (evolutionary development or "family tree"). This claim is still found in most high school and college texts although most embryologists now believe this theory to be completely discredited.

Over 40 years ago Waldo Shumway of the University of Illinois said, with respect to the theory of embryological recapitulation (also called the "biogenetic law"), that a consideration of the results of experimental embryology "seem to demand that the hypothesis be abandoned."[30] Walter J. Bock of the Department of Biological Sciences of Columbia University says:

> . . . the biogenetic law has become so deeply rooted in biological thought that it cannot be weeded out in spite of its having been demonstrated to be wrong by numerous subsequent scholars.[31]

Many similar quotes to this effect may be cited (see for example the excellent section by Davidheiser on the theory of embryological recapitulation[32]).

One of the more popular ideas expressed by those who believe in embryological recapitulation is the idea that the human embryo (as well as the embryos of all mammals, reptiles, and birds) has "gill slits" during early stages of its development. The human embryo does have a series of bars and grooves in the neck region, called pharyngeal pouches, which superficially resemble a series of bars and grooves in the neck region of the fish which do develop into gills. In the human, however (and in other mammals, birds, and reptiles), these pharyngeal pouches do not open into the throat (they thus cannot be "slits"), and they do not develop into gills or respiratory tissue (and so they cannot be "gills"). If they are neither gills or slits, how then can they be called "gill-slits"? These structures actually develop into various glands, the lower jaw, and structures in the inner ear.

If the human embryo recapitulates its assumed evolutionary ancestry, the human heart should begin with one chamber and then develop successively into two, then three, and finally four chambers. Instead the human heart begins as a two-chambered organ which fuses to a single chamber which then develops directly into four chambers. In other words, the sequence is 2-1-4, not 1-2-3-4 as required by the theory. The human brain develops before the nerve cords and the heart before the blood vessels, both out of the assumed evolutionary sequence. It is because of many similar contradictions and omissions that the theory of embryological recapitulation has been abandoned by embryologists.

Evolutionists at one time listed about 180 organs in the human body considered to be no more than useless vestiges of organs that were useful in man's animal ancestors. With increasing knowledge, however, this list has steadily shrunk until the number has been reduced to practically zero. Important organs such as the thymus gland, the pineal gland, the tonsils, and the coccyx (tail bone) were once considered vestigial. The thymus gland and the tonsils are involved in defense against disease. The appendix contains tissue similar to that found in the tonsils and is also active in the fight against foreign invaders. The coccyx is not a useless vestige of a tail, but serves an important function as the anchor for certain pelvic muscles. Furthermore, one cannot sit comfortably following removal of the coccyx.

Evolutionists cite the fact that many different kinds of animals have structures, organs (called homologous structures), and metabolism that are similar. That this is true is quite evident. Is it surprising that the biochemistry (life chemistry or metabolism) of the human is very similar to that of a rat? After all, don't we eat the same kind of food, drink the same water, and breathe the same air? *If* evolution were true, similarities in structure and metabolism would be a valuable aid in tracing evolutionary ancestries, but it is worthless as evidence *for* evolution. These types of similarities are predicted by both the creation and the evolution models. Such similarities are actually the result of the fact that creation is based upon the master plan of the Master Planner. Where similar functions were needed, the Creator used similar structures and life chemistry to perform these functions, merely modifying these structures and metabolic pathways to meet the individual requirement of each organism.

Much of the morphological and genetic evidence related to homologous structures in fact directly contradicts predictions based on evolution theory. Much of this contradictory data is discussed by Sir Gavin de Beer, a firm advocate of evolution theory, in his Oxford Biology Reader entitled "Homology, An Unsolved Problem."[33] Sir Gavin chose this title because the evidence is contradictory to what he, as an evolutionist, would expect.

After citing much of this contradictory evidence, Sir Gavin mentions the cruelest blow of all — the contradiction between the genetic data and the concept of inheritance of homologous structures from a common ancestor. After some discussion Sir Gavin says:

> It is now clear that the pride with which it was assumed that the inheritance of homologous structures from a common ancestor explained homology was misplaced; for such inheritance cannot be ascribed to identity of genes. The attempt

> to find "homologous" genes, except in closely related species, has been given up as hopeless.

If homologous structures exist because animals (or plants) which possess these similar structures do so because they have inherited these homologous structures through evolution from a common ancestor which possessed the structure, then certainly these creatures should share in common the genes each inherited from the common ancestor which determined the homologous structure. In other words, the set of genes in each one of these creatures which determines the homologous structure should be nearly identical (thus "homologous"). But this is not the case. When the homologous structure is traced back to the genes which determine it, these genes are found to be completely different in the animals (or plants) possessing the homologous structure.

Evolutionists believe that structures change (or evolve) because genes change (or evolve). Thus, if genes change, certainly the structure or function governed by these genes should change. Conversely, if the structure or function has remained unchanged, then the genes governing this structure or function would remain unchanged. These are clearly the predictions that would be made if evolution were true. The actual genetic data, however, directly contradicts these predictions.

Because of this fact, evolutionists are forced to postulate an incredible situation. Thus, as cited by Sir Gavin, S. C. Harland has stated:

> The genes, as a manifestation of which the character develops, must be continually changing . . . . we are able to see how organs such as the eye, which are

> common to all vertebrate animals, preserve their essential similarity in structure or function, though the genes responsible for the organ must have become wholly altered during the evolutionary process.[34]

What an incredible suggestion! Genes, for example those governing the eyes, evolve into entirely different genes, but the structure (the eye) governed by these genes remains unchanged! In their attempt to resolve the contradictions between the genetic data and evolution theory, evolutionists are forced to postulate the most preposterous hypotheses. No naturalistic, mechanistic process could accomplish such an amazing physical arrangement — the structures being nearly identical, but the genes being completely different. The evidence certainly indicates that the genetic engineer that brought about such an incredible arrangement was an omnipotent Creator.

Although Sir Gavin can think of no alternative to the suggestion of Harland, he evidently feels very uncomfortable about it, for he says:

> But if it is true that through the genetic code, genes code for enzymes that synthesize proteins which are responsible (in a manner still unknown in embryology) for the differentiation of the various parts in their normal manner, what mechanism can it be that results in the production of homologous organs, the same "patterns," in spite of their *not* being controlled by the same genes? I asked that question in 1938, and it has not been answered.[35] (Emphasis added.)

It has not been answered because no answer is

available that is compatible with evolution theory. It is highly recommended that those interested in the creation/evolution question obtain copies of both the Oxford Biology Reader and the 1938 publication by Sir Gavin. The 1938 publication discusses both homology and embryology and the problems these generate for evolution theory.[36]

The suggestion of Harland that structures can remain unchanged while the genes governing them become completely altered, in addition to the contradiction to evolution theory mentioned above, contradicts another basic assumption of evolution theory — that is, that evolution occurs through natural selection. In this case it is obvious that, while the genes have become wholly altered, and thus have evolved drastically (according to evolutionists), natural selection could not possibly have been involved since the structure (in this case the eye) remains unchanged.

Natural selection, according to evolution theory, involves an interaction between the environment and the structures and functions (the phenotype) of plants and animals. There is no way the genes (the genotype) can be involved in this interaction without involving the phenotype. That is, the genotype can become involved only by its effect on the phenotype. If this is the case, then in the supposed evolutionary transformation of the genes suggested above by Harland, how could natural selection preserve and enhance the precentage of the mutant versus the original or unchanged variety throughout the series of changes required, since the structure itself remains unchanged? Obviously, natural selection is excluded. Furthermore, this has occurred numerous times, according to Sir Gavin, because, as he says, the attempt to find "homologous" genes, except in closely related species (thus all derived from a single created kind according to creationists) has been given up as hopeless. Repeatedly, then, according to evolution theory, genes have become wholly altered with no change in the structure or function governed by these genes, the process thus being wholly independent of natural selection, the supposed driving force of evolution!

CONCLUSION

Kerkut, although not a creationist, authored a notable little volume to expose the weaknesses and fallacies in the usual evidence used to support evolution theory. In the concluding paragraph of this book, Kerkut stated that:

> . . . there is the theory that all the living forms in the world have arisen from a single source which itself came from an inorganic form. This theory can be called the "General Theory of Evolution" and *the evidence that supports it is not sufficiently strong to allow us to consider it as anything more than a working hypothesis.*[37] (Emphasis added.)

There is a world of difference, of course, between a working hypothesis and established scientific fact. The "fact of evolution" is actually the *faith* of evolutionists in their particular world view.

No less a convinced evolutionist than Thomas H. Huxley acknowleged that:

> . . . "creation," in the ordinary sense of the word is perfectly conceivable. I find

> no difficulty in conceiving that, at some former period, this universe was not in existence, and that it made its appearance in six days (or instantaneously, if that is preferred), in consequence of the volition of some preexisting Being. Then, as now, the so-called *a priori* arguments against Theism and, given a Deity, against the possibility of creative acts, appeared to me to be devoid of reasonable foundation.[38]

The refusal of the establishment within scientific and educational circles to consider creation as an alternative to evolution is thus based above all on the insistence upon a purely atheistic, materialistic, and mechanistic explanation for origins to the exclusion of an explanation based on theism. Restricting the teaching concerning origins to this one particular view thus constitutes indoctrination in a religious philosophy. Constitutional guarantees of separation of church and state are violated and true science is shackled in dogma.

After many years of intense study of the problem of origins from a scientific viewpoint, I am convinced that the facts of science declare special creation to be the only *logical* explanation of origins.

REFERENCES

1. G. G. Simpson, *Tempo and Mode in Evolution*, Columbia University Press, New York, p. 105 (1944).
2. G. G. Simpson, *The Meaning of Evolution*, Yale University Press, New Haven, 1949, p. 231.
3. G. G. Simpson, in *The Evolution of Life*, Sol Tax, ed., University of Chicago Press, Chicago, 1960, p. 149.
4. T. Dobzhansky, *Science*, Vol. 175, p. 49 (1972).
5. G. G. Simpson, *The Major Features of Evolution*, Columbia University Press, New York, 1953, p. 360.
6. Ref. 5, pp. 360-376.
7. Ref. 1, pp. 105-124.
8. Ref. 3, pp. 149-152.
9. W. Thorpe, *New Scientist*, Vol. 43, p. 635 (1969).
10. C. A. Arnold, *An Introduction to Paleobotany*, McGraw-Hill Pub. Co., New York, 1947, p. 7.
11. E. J. H. Corner, in *Contemporary Botanical Thought*, A. M. MacLeod and L. S. Cobley, eds., Quadrangle Books, Chicago, 1961, p. 97.
12. E. C. Olson, *The Evolution of Life*, The New American Library, New York, 1965, p. 94.
13. N. D. Newell, *Proc. Amer. Phil. Soc.*, April, 1959, p. 267.

14. D. M. Raup and S. M. Stanley, *Principles of Paleontology*, W. H. Freeman and Co., San Francisco, 1971, p. 306.
15. L. du Nouy, *Human Destiny*, The New American Library, New York, 1947, p. 63.
16. O. Kuhn, *Acta Biotheoretica*, Vol. 6, p. 55, (1942).
17. A. H. Clark, in *The New Evolution: Zoogenesis*, A. H. Clark, ed., Williams and Wilkins, Baltimore, 1930, p. 189.
18. T. Dobzhansky, M. K. Hecht, and W. C. Steere, eds., *Evolutionary Biology*, Volume 6, Apleton-Century-Crafts, New York, 1972.
19. L. Van Valen, *Science,* Vol. 180, p. 488 (1973).
20. L. du Nouy, Ref. 15, p. 74.
21. R. B. Goldschmidt, *American Scientist*, Vol. 40, p. 97 (1952).
22. F. W. Cousins, *Creation Research Society Quarterly*, Vol. 8, p. 99 (1971).
23. R. B. Goldschmidt, *The Material Basis of Evolution*, Yale University Press, New Haven, 1940.
24. R. B. Goldschmidt, Ref. 21, pp. 84-98.
25. S. J. Gould, *Natural History*, Vol. 86, pp. 22-30 (1977).
26. S. J. Gould, *ibid.*, Vol. 86, p. 13 (1977).
27. S. J. Gould, *ibid.*, Vol. 85, p. 37 (1976).
28. D. B. Kitts, *Evolution*, Vol. 28, p. 467 (1974).
29. N. Macbeth, *American Biology Teacher*, November 1976, p. 495.
30. W. Shumway, *Quarterly Review of Biology*, Vol. 7, p. 98 (1932).
31. W. J. Bock, *Science*, Vol. 164, p. 684 (1969).
32. B. Davidheiser, *Evolution and Christian Faith*, Presbyterian and Reformed Publ. Col., Philadelphia, 1969, p. 240.

33. G. R. de Beer, *Homology, An Unsolved Problem*, Oxford University Press, Oxford, 1971.
34. Ref. 33, p. 16; S. C. Harland, *Biological Reviews*, Vol. 11, p. 83 (1936).
35. Ref. 33, p. 16.
36. G. R. de Beer, in *Evolution: Essays Presented to E. S. Goodrich*, G. R. de Beer, Ed., Clarendon Press, Oxford, 1938.
37. G. A. Kerkut, *Implications of Evolution*, Pergamon Press, New York, 1960, p. 157.
38. T. H. Huxley, quoted in *Life and Letters of Thomas Henry Huxley*, Vol. I, L. Huxley, ed., Macmillan, p. 241, 1903.

SUBJECT INDEX

AUTHOR INDEX

INDEX OF ILLUSTRATIONS

LIBRARY
OF
MOUNT ST. MARY'S
COLLEGE

definitions runs throughout the book, which also includes an extensive bibliography. 8½″ x 11″.

No. 114, Paper $3.25

Public School Teacher's Guide, No. 116, Paper $1.50

Overhead Transparencies, No. 118, $17.00

Origins Set (one each of above) No. 120, $20.00

DINOSAURS: Those Terrible Lizards

Duane T. Gish, Ph.D.

A book for young people which discusses the dinosaur question from a creationist perspective. Did people in earlier times live with dinosaurs? Were dragons of ancient legends really dinosaurs? Does the Bible speak about dinosaurs? All these questions are addressed in this book by noted scientist, Dr. Gish, author of the bestseller, *Evolution? The Fossils Say NO!* Illustrated profusely in color, 9″ x 11″.

No. 046, Cloth $5.95

Technical Monograph No. 1

Speculations and Experiments Related to the Origin of Life (A Critique)

Duane T. Gish, Ph.D.

An analysis and devastating critique of current theories and laboratory experiments attempting to support a naturalistic development of life from nonliving chemicals. No. 158, Paper $3.50

Technical Monograph No. 2

Critique of Radiometric Dating

Harold Slusher, M.S., D. Sc.

Evaluation and refutation, from sound principles of physics, of the most important radiometric methods of determining geologic ages. No. 159, Paper $3.50

Technical Monograph No. 4

Origin and Destiny of the Earth's Magnetic Field

Thomas G. Barnes, M.S., Sc.D.

A technical exposition of one of the most conclusive, unanswerable proofs that the earth is less than 10,000 years old, as calculated from the rapidly-decaying magnetic field of the earth.

No. 161, Paper $3.50

Streams of Civilization

Albert Hyma, Ph.D. and Mary Stanton, Ed.D.

Ancient History to 1572 A.D. A world history text for junior and senior high schools. An all-inclusive, value-oriented presentation, blending archaeology, biology, philosophy, science, and the fine arts to clarify history. Creation and evolution are investigated as unbiased, scientific approaches to the origin of man and his universe. Over 350 illustrations, maps, and photos, with a quick reference comprehensive index. 8½ " x 11 ", 411 pages.

No. 145, Cloth $12.95
Teacher's Guide, No. 146, Paper $ 2.50

BIOLOGY: A Search for Order in Complexity

Ed. by John N. Moore, Ed.D. & Harold S. Slusher, M.S., D.Sc.

The high school biology text presenting a unique, scientific approach to origins. Evolution and special creation are both presented as viable alternatives. Highly readable and teachable, the text is designed for two semesters and has three manuals for teachers and students to assist in learning the principles of biology. Can be adapted for either junior high or junior college. Contains 596 pages, 436 illustrations, extensive bibliographies and glossaries, plus exhaustive indexes.

Textbook, No. 025, Cloth $9.95
Teacher's Guide, No. 026, Paper $1.95
Student Lab Manual, No. 027, Paper $2.95
Teacher Lab Manual, No. 028, Paper $2.95